The Power of the Mine

The Power of the Mine

A Transformative Opportunity for Sub-Saharan Africa

Sudeshna Ghosh Banerjee, Zayra Romo, Gary McMahon,
Perrine Toledano, Peter Robinson, and Inés Pérez Arroyo

WORLD BANK GROUP

SAFETE
South African Fund for African Energy, Transport and Extractive Industries

ESMAP
Energy Sector Management Assistance Program

Contents

Boxes

Figures

Tables

Foreword

A Potential Game Changer for Sub-Saharan Africa

Africa is blessed with energy resources yet they go largely untapped. As a result, only one in three Africans has access to energy, which stymies economic growth on the continent and seriously limits human potential and well-being. If nothing changes, Sub-Saharan Africa as a region will actually see the number of people without electricity increase from 590 million in 2013 to 655 million by 2030.

Against this backdrop, *The Power of the Mine: A Transformative Opportunity for Sub-Saharan Africa* (http://www.worldbank.org/africa/powerofthemine) tackles a fairly new area of inquiry on the energy front: how can an energy-intensive, high-volume customer like the mining industry improve its contribution to energy supply, help expand access, and attract private capital into the space?

The report shows that the mining industry in Sub-Saharan Africa has been sourcing power in innovative ways—some involving national utilities and some not. The self-supply arrangement imposes a loss for everyone—people, utilities, mines, and national economies. Since 2000, mines in Africa have spent around $15.3 billion to cover their own electricity investment and operating costs and have installed 1,590 megawatts of generating capacity. None of this power made it onto a national grid.

Irrespective of the country—ranging from areas where a grid barely exists, forcing mines to secure their own generation, to those with large, integrated grid systems—there is great potential for the mining industry to be used as an "anchor customer" to unlock energy resources for the sustainable development of the power sector.

While the economic and business case for power–mining integration is strong, the report shows that this opportunity has largely been undeveloped. In some countries, integration could help connect many customers to mini-grids or national grids. In others, it has a facilitating role to support the power sector through greater mobilization of revenues from energy sales.

The report also points to the challenges that must be overcome in this new and—in a developing country context—relatively unchartered area. But these are not insurmountable, and many countries that have integrated mining demand successfully in their power sectors offer proof that this untapped potential can be harnessed for national development.

African policymakers can put in place the risk-mitigation mechanisms to create an operating environment where innovative power–mining arrangements can thrive. And the World Bank Group can use its convening power to generate a dialogue around the topic, share knowledge, and provide an array of financial instruments to achieve transformational impact. Through this report, we seek to highlight not only the challenges facing energy sector development in Sub-Saharan Africa, but also the new emerging opportunities to overcome them for a better future for all.

Anita Marangoly George
Senior Director,
Energy and Extractives Global Practice
The World Bank Group

Acknowledgments

This report was drafted by a team comprising Sudeshna Ghosh Banerjee, Zayra Romo, and Gary McMahon and supported by Inés Pérez Arroyo, Arsh Sharma, Joeri de Wit, and Aly Sanoh. The work was carried out under the guidance of Lucio Monari, Meike van Ginneken, Vivien Foster, Christopher Sheldon, and Jamal Saghir. Substantive inputs were received from Bastiaan Verink, Richard MacGeorge, and Pankaj Gupta.

This report draws from two background papers produced by the Columbia Center on Sustainable International Investment (CCSI) at Columbia University and Economic Consulting Associates. The team from the CCSI was led by Perrine Toledano and comprised Kwabena Okrah and Angelo Antolino. The Economic Consulting Associates team was led by Peter Robinson and included Grigorios Varympopiotis, Fred Beelitz, and Erin Boyd.

The team is grateful to colleagues in the Energy and Extractives Practice, who contributed valuable insights and background material, particularly in the eight countries that have been studied at length in this work. Specifically, the team benefited tremendously from discussions with Moez Cherif, Sunil W. Mathrani, Waqar Haider, Bassem Abou-Nehme, Rob Mills, Robert Schlotterer, Bobak Rezaian, Brent Hampton, Daniel Murphy, Philippe Durand, David John Santley, Boubacar Bocoum, Morten Larsen, Natalia Cherevatova, Ekaterina Mikhaylova, Brigitte Bocoum, Remy Pelon, Kristina Svensson, Sarwat Hussain, and Pedro Antman.

The team thanks peer reviewers Jan Walliser, Wendy Hughes, Johannes Herderschee, Alan Moody, Sebastian Dessus, Mohinder Gulati, Masami Kojima, Marcelino Madrigal, and Bryan Land, who provided valuable comments and constructive insights at various stages of this work.

Bruce Ross-Larson and Jack Harlow of Communications Development Incorporated and Norma Adams edited the report.

Financial support from the Energy Sector Management Assistance Program, the World Bank Budget, and South Africa Fund for Transportation and Energy is gratefully acknowledged.

About the Authors

Sudeshna Ghosh Banerjee is a Senior Economist in the Global Energy and Extractives Practice at the World Bank Group. She has worked on energy and infrastructure issues in the South Asia and Africa departments in both operations and analytic assignments. She focuses on project economics, monitoring and evaluation, and on a broad range of energy sector issues including energy access, energy subsidies, renewable energy, and sector assessments. She holds a PhD in Public Policy from the University of North Carolina at Chapel Hill and MA and BA degrees in Economics from Delhi University.

Zayra Romo is a Senior Energy Specialist in the Global Energy and Extractives Practice at the World Bank Group. She has led technical analysis and been involved in investment operations related to energy infrastructure projects in Latin America and Sub-Saharan Africa. In her current role, she leads the design and implementation of several investment operations in generation, transmission, and distribution in Africa. She has assisted client countries in power sector planning and development of renewable energy programs, primarily in Liberia and Tanzania. Prior to joining the World Bank, she worked for Electricite de France in Mexico as a technical power analyst. She has a degree in Electrical and Electronic Engineering from the National Autonomous University of Mexico (UNAM), and an MSc in Energy Conversion and Management from the University of Offenburg, Germany.

Gary McMahon is a Senior Mining Specialist in the Global Energy and Extractives Practice at the World Bank Group. He has worked on issues in the mining sector for over 20 years, particularly in the areas of local benefits, social and environmental impacts, and macroeconomic opportunities and risks. His most recent publication examines the socioeconomic and human development impacts of mining since the turn of the century in mineral-rich low- and lower-middle-income countries. Previously, he examined the history of mining sector technical assistance by the World Bank, "World Bank Support for Mining Sector Reform: An Evolutionary Approach." With a PhD in Economics from the University of Western Ontario, he has written extensively on development issues on a wide variety of subjects and has edited nine volumes, ranging from the impacts of large mining operations on community development to the political economy of market reform in developing countries. He currently works on mining issues in Zimbabwe, Guinea-Bissau, Bolivia, Armenia, Romania, the Kyrgyz Republic, Tajikistan, and Kazakhstan.

Perrine Toledano is the Senior Economics Researcher of the Columbia Center on Sustainable International Investment (CCSI); she also leads the CCSI's focus on Extractive Industries and Sustainable Development. In those capacities, she leads research, training, and advisory projects on fiscal regimes; financial modeling; leveraging of extractive industry investments in rail, port, telecommunications, water, and energy infrastructure for broader development needs; local content; revenue management; and optimal legal provisions for development benefits. Prior to joining the CCSI, she worked as a consultant for several nonprofit organizations (World Bank, U.K. Department for International Development, Revenue Watch Institute) and private sector companies (Natixis Corporate Investment Bank, Ernst and Young). Her experience includes auditing, financial analysis, IT for capital markets, public policy evaluation, and cross-border project management. She has a an MBA from ESSEC (Paris, France) and a Masters of Public Administration from Columbia University.

Peter Robinson is an economist who spent the first 25 years of his career based in Zimbabwe, working on development issues in East and Southern Africa. Since 2007, he has been Economic Consulting Associates' Director for the Africa region, building up the energy and water practice in Africa, as well as working in other parts of the world. His work has spanned policy and strategy, sectoral reform and institutional analysis, regulation, tariff studies, and service delivery. He has an academic background in electrical engineering and economics and a PhD in engineering-economic systems. He has a strong interest in the relationship of infrastructure to overall economic development, with regional strategies set to play a particularly important role in Sub-Saharan Africa. In recent years, he has increasingly focused on ways in which low income households can access and use modern infrastructural services on an affordable and sustainable basis.

Inés Pérez Arroyo is a Consultant in the Global Energy and Extractives Practice at the World Bank Group. She has been involved in the preparation and supervision of several investment operations in the energy sector in West Africa, and has participated in the analytical and desk research work for various studies in the power sector in Sub-Saharan Africa. Before joining the World Bank Group, she worked at the Economic and Commercial Office of the Embassy of Spain in Tokyo as an international trade consultant, where she was in charge of market research and analysis of the Japanese and Spanish energy sectors, as well as support private investments of Spanish companies in Japan. She holds a degree in Physics from the University of Valladolid (Spain) and a Master in International Business Management from the International University Menéndez Pelayo (Spain).

Abbreviations

ARSEL	Agence de Régulation du Secteur de l'Electricité (Cameroon)
CAGR	compound annual growth rate
CAPP	Central African Power Pool
CAR	Central African Republic
CCGT	combined-cycle gas turbine
CEC	Copperbelt Energy Corporation
CNELEC	Conselho Nacional de Electricidade (Mozambique)
CO_2	carbon dioxide
CSR	corporate social responsibility
ECG	Electricity Company of Ghana
EDG	Electricité de Guinée
EDM	Electricidade de Moçambique
ERB	Energy Regulatory Board (Zambia)
EWURA	Energy and Water Utilities Regulatory Authority (Tanzania)
GDP	gross domestic product
HCB	Hydropower plant Cahora Bassa (Mozambique)
HDI	human development index
HFO	heavy fuel oil
IMF	International Monetary Fund
IPP	independent power producer
LNG	liquefied natural gas
LRMC	long-run marginal cost
NWEC	Northwest Energy Company (Zambia)
O&M	operating and maintenance
PGM	platinum group metals
PPA	power purchase agreement
PPP	public–private partnership
PURC	Public Utilities Regulatory Commission (Ghana)

SAPP	Southern African Power Pool
SNEL	Société Nationale d'Electricité (Democratic Republic of Congo)
SNIM	Société Nationale Industrielle et Minière (Mauritania)
SOMELEC	Société Mauritanienne d'Electricité
SSA	Sub-Saharan Africa
STE	Sociedade Nacional de Transporte de Energía
TANESCO	Tanzania Electric Supply Company
UG	underground
VALCO	Volta Aluminum Company Limited
VRA	Volta River Authority
ZESA	Zimbabwe Electricity Supply Authority
ZESCO	Zambia Electricity Supply Corporation

Units of Measure

Bcf	billion cubic feet
cfr	cost and freight
c/kWh	cents per kilowatt-hour
GJ	gigajoule
GW	gigawatt
GWh	gigawatt-hour
kg	kilogram
km	kilometer
kV	kilovolt
kWh	kilowatt-hour
kWh/t	kilowatt hour/ton
m	meter
mt	metric ton
MW	megawatt
MWh	megawatt-hour
tr oz	troy ounce

All dollar amounts are U.S. dollars unless otherwise indicated.

Overview

Providing households and firms in Sub-Saharan Africa (SSA) with adequate and reliable electricity in a sustainable manner—a critical challenge—is important to eradicate extreme poverty and boost shared prosperity. SSA's power sector lags behind that of all other regions, with an installed generating capacity of only 80 gigawatts (GW). More than half of that amount (45 GW) is generated by South Africa, followed far behind by Nigeria at 6 GW. Close to half of SSA firms identify unreliable electricity as one of the biggest constraints to doing business. Pervasive outages cost about 5 percent in lost annual sales. About 44 percent of firms cope by owning or sharing a generator, fueled typically by diesel or heavy fuel oil (HFO). Two-thirds of Africans depend on kerosene or dry-cell batteries to meet their subsistence power needs, the costs of which often burden household budgets and human and environmental health.

Business as usual is not even remotely good enough. SSA's installed generating capacity has risen 1.7 percent a year over the past 20 years. Long-term demand projections point to a need to raise installed capacity to 700 GW by 2040 to meet the needs of economic growth, suppressed demand, and increased access to electricity (Eberhard and others 2011).

This suboptimal situation exists amid vast energy resources, most of which are concentrated in a handful of countries. Geothermal potential of about 17 GW is present in six countries along East Africa's Rift Valley. Hydropower, coal, and natural gas are the region's most abundantly available resources. Only 8 percent of its hydro potential (about 400 GW)—compared with 18 percent for Latin America and 20 percent for Asia—has been developed (Kumar and others 2011). SSA's gas resource base, a third of which is constituted by new finds in Mozambique and Tanzania, is estimated at 329 trillion cubic feet, more than twice the proved reserve figure. This quantity of gas could generate 100 GW of power a year for more than 70 years if combined-cycle gas-turbine (CCGT) power plants are considered (Santley, Schlotterer, and Eberhard 2014).

Mines, with their substantial and growing need for power, can help harness these energy resources as anchor consumers. Anchor consumers are high-volume customers that provide a captive source of demand and a consistent source of revenues. This is not a new concept: numerous industrial mills and factories served as anchor loads in earlier stages of development in today's rich economies. Now this concept is being applied in developing countries—cellphone towers, for example. Mines could similarly serve as anchor consumers because large amounts of power are required to run their processes—power rarely constitutes less than 10 percent of mining's operating costs and often rises above 25 percent. The boom in mineral prices since 2003 has also translated into rising demand for power.

SSA hosts a substantial portion of known global mineral reserves. In 2011 the region produced 22 percent of gold, 58 percent of cobalt, 7 percent of copper, 95 percent of platinum group metals (PGM), and 18 percent of uranium. SSA is also becoming an important region for iron ore and coal following important recent discoveries. Mining is one of SSA's most important industries in its contribution to exports, fiscal revenues, and gross domestic product (GDP), accounting for more than half of total exports in Burkina Faso, the Democratic Republic of Congo, Guinea, Mauritania, Mozambique, and Zambia. Fiscal revenues from mining account for more than 20 percent of the total in Botswana, the Democratic Republic of Congo, and Guinea. The projected mining investment for 2013–20 was up to 75 billion; in Guinea, Liberia, and Sierra Leone, it could represent an overwhelming proportion of GDP (2012) (figure O.1). And mining's potential contribution to African economies has yet to be realized. Until this decade, SSA severely lagged behind the rest of the world in exploration spending. But over 2002–12, Africa's share of global mining exploration—itself steeply rising—climbed from 10 percent to 17 percent. Africa's absolute

Figure O.1 Mining Investment, Current and Forecast, 2000–20

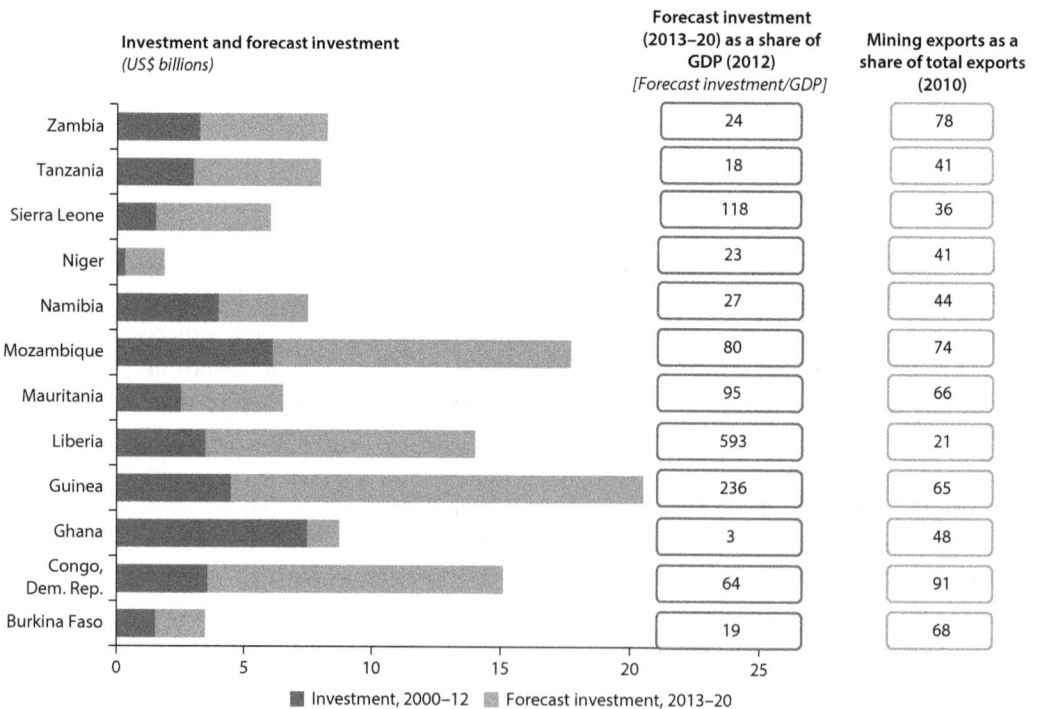

	Investment and forecast investment (US$ billions)	Forecast investment (2013–20) as a share of GDP (2012) [Forecast investment/GDP]	Mining exports as a share of total exports (2010)
Zambia		24	78
Tanzania		18	41
Sierra Leone		118	36
Niger		23	41
Namibia		27	44
Mozambique		80	74
Mauritania		95	66
Liberia		593	21
Guinea		236	65
Ghana		3	48
Congo, Dem. Rep.		64	91
Burkina Faso		19	68

■ Investment, 2000–12   ■ Forecast investment, 2013–20

Source: Based on World Bank 2012.
Note: The investment figures are averages across a range. For Mozambique, forecast investment excludes natural gas.

spending on exploration rose more than 700 percent, reaching $3.1 billion in 2012 (Wilburn and Stanley 2013).

Power–mining integration can create a win-win situation. Mining companies could be anchor customers for utilities, facilitating generation and transmission investments by producing the economies of scale needed for large infrastructure projects, in turn benefiting all consumers. Utilities can also secure large revenues from creditworthy customers. Grid supply in turn costs mines less than self-supply from diesel and HFO and allows the mines to focus on their core business. Such arrangements could raise the GDP of host countries by allowing for more mineral beneficiation and providing more job opportunities in local firms selling goods and services to the mines, whose competitiveness would be greatly enhanced by lower power costs. Leveraging mining's demand for power and its capital investments in power infrastructure thus offers a transformational opportunity to develop the power sector of Africa's mineral-rich economies.

This study is the first to systematically analyze the potential and challenges of power–mining integration in SSA. The following analytic tools were used: the power demand from mining was projected to 2020; a typology of mining power-sourcing arrangements was created, highlighting intermediate options between self-supply and grid supply; power–mining integration scenarios were simulated for Guinea, Mauritania, Mozambique, and Tanzania, reporting substantial cost savings from shared infrastructure, potential for regional integration, and expansion of electrification; and lessons were learned from experience in Cameroon, the Democratic Republic of Congo, Ghana, and Zambia, where mining and power have achieved some integration and provide a reference for identifying benefits and risks. Underpinning the Africa-wide analysis is the Africa Power–Mining Database 2014 created for this study (box O.1).

Box O.1 Africa Power–Mining Database 2014—and Two Probability Scenarios

The Africa Power–Mining Database 2014 (found at http://www.worldbank.org/africa /powerofthemine) draws on basic mining data from Infomine surveys, the United States Geological Survey, annual reports, technical reports, feasibility studies, investor presentations, sustainability reports on property-owner websites or filed in public domains, and mining websites (Mining Weekly, Mining Journal, Mbendi, Mining-technology, and Miningmx).

Comprising 455 projects in 28 SSA countries with each project's ore reserve value assessed at more than $250 million, the database collates publicly available and proprietary information. It also provides a panoramic view of projects operating in 2000–12 and anticipated demand in 2020. Demand in 2020 can speculatively be considered a lower-bound estimate of demand in 2025 or 2030. The analysis is presented over three timeframes: pre-2000, 2001–12, and 2020 (each containing the projects from the previous period except for those closing during that previous period).

box continues next page

Mining Demand for Power Is Expected to Triple by 2020 from 2000

Mining demand for power will be substantial, potentially reaching 23,443 megawatts (MW) by 2020. SSA's mining sector required about 7,995 MW in 2000 and more than 15,124 MW in 2012 (figure O.2a). The power demand in 2020 is envisaged at 142–155 percent of 2012 demand, given the two probability scenarios—even more impressive because a large part of the demand is anticipated to emerge from outside South Africa. While South Africa is projected to add sizable mining demand for power and grow at 3.5 percent a year between 2012 and 2020, growth in other SSA countries, projected at 9.2 percent, is yet more impressive if all projects in the two probability scenarios are realized.

Demand will come overwhelmingly from Southern Africa, dominated by South Africa. By far SSA's most important mining country, South Africa accounted for 70 percent of mining power demand in 2000 and 66 percent in 2012. But its contribution is forecast to decline to 56 percent in 2020. Even without South Africa, Southern Africa will have the highest demand, owing largely to the substantial requirements in Mozambique and Zambia, followed by Central Africa and West Africa (figure O.2b).

Power demand has been concentrated in a small group of metals: aluminum leads, followed by copper, PGM, chromium, and gold. But for aluminum, there has been little new activity since 2000, and power demand will grow by 2020 only if low-probability projects are realized. Copper and platinum will provide the largest increased demand, at 2,150 MW and 2,010 MW, respectively, when high- and low-probability projects are included. Smelting, the most power-intensive part of the cycle, contributes about three-quarters of power demand, along with refinement processes. Separation and crushing processes, which are less power-intensive, are envisaged to grow (starting from a low base) by 9 percent and 8 percent a year, respectively, over 2012–20.

Mining Demand for Power in 2020 Can Dominate Other Sectors in a Few Countries

In some countries, mining power demands overwhelm power demand from other sectors. In 2012 it accounted for 24 percent of SSA's domestic demand and is expected to rise to 30 percent by 2020, when high- and low-probability

Figure O.2 Mining Demand for Power

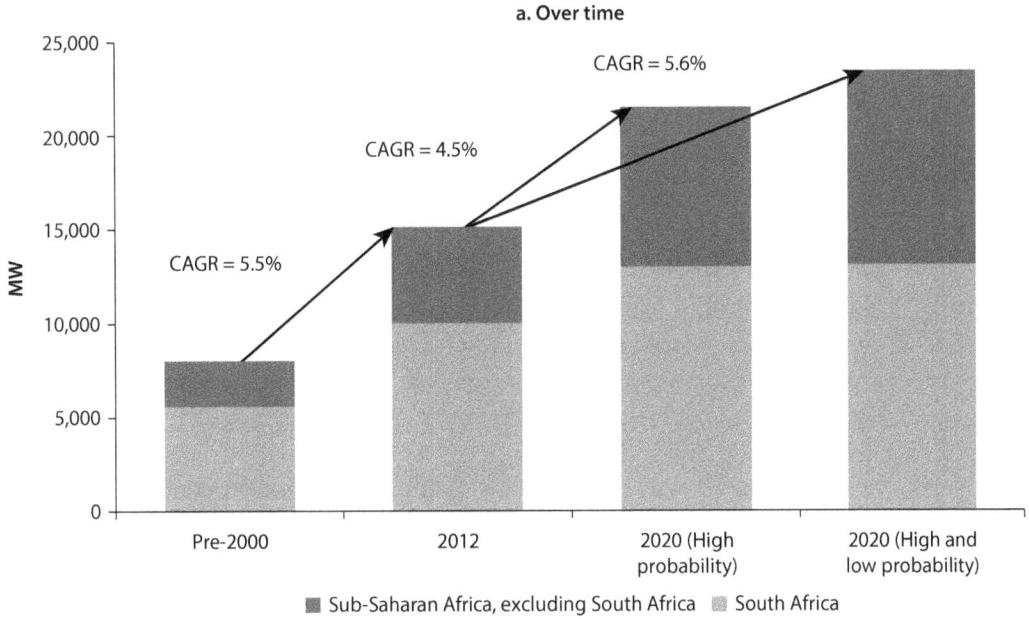

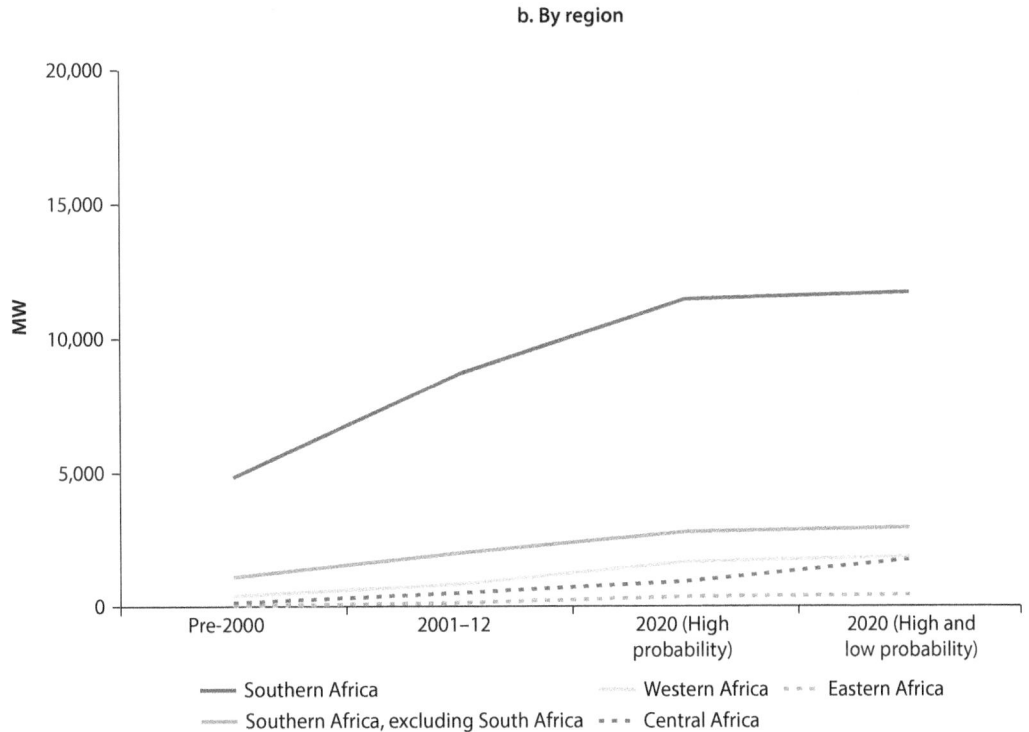

a. Over time

CAGR = 5.6%

CAGR = 4.5%

CAGR = 5.5%

■ Sub-Saharan Africa, excluding South Africa ■ South Africa

b. By region

— Southern Africa   Western Africa - - - Eastern Africa
— Southern Africa, excluding South Africa - - - Central Africa

Source: Africa Power–Mining Database 2014, World Bank, Washington, DC.
Note: CAGR = compound annual growth rate.

projects are considered (figure O.3a). In Guinea, Liberia, and Mozambique, mining demand is expected to exceed total nonmining demand by 2020 (figure O.3b). For many other countries, mining will account for a substantial proportion of total demand in 2020. Compared with available supply, mining demand can impose a substantial burden. If grid supply remains at 2012 levels, mining demand could account for up to 35 percent of grid supply by 2020.

The Future Is Frontier Action in the Dynamic Space between Self-Supply and Grid Supply

Traditionally, mines have sourced power from the grid, depending on an adequate, reliable supply. In the other prominent arrangement, self-supply, a mine produces its own power because of the high costs of extending transmission and distribution networks to the mining site, low security of grid-supplied electricity, and high tariffs where the grid is powered by expensive fuel sources. Between self-supply and grid supply are at least six intermediate arrangements: self-supply and corporate social responsibility (CSR), self-supply and sell to the grid, grid supply and self-supply backup, mines sell collectively to the grid, mines invest in the grid, and mines serve as anchor demand for an independent power producer (IPP) (table O.1). The most common intermediate options over 2000–20 are "self-supply and sell to the grid" and "mines invest in the grid." Combinations of arrangements—or transitional arrangements—can evolve over the mine's lifetime.

Figure O.3 Comparison of Mining and Nonmining Demand for Power

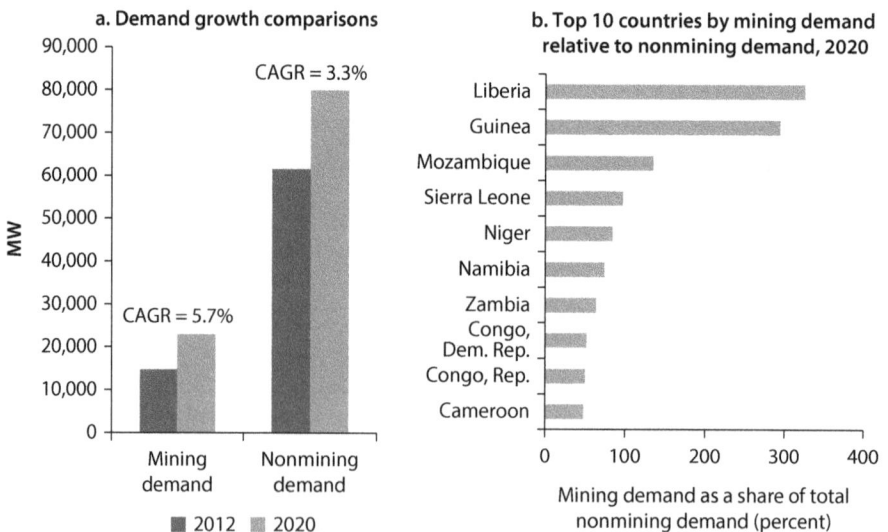

Source: Africa Power–Mining Database 2014, World Bank, Washington, DC.
Note: Panel a includes 27 countries for which both mining demand and total demand figures are available. CAGR = compound annual growth rate.

Table O.1 Typology of Power-Sourcing Arrangements

	Self-supply	Self-supply + CSR	Intermediate options					Grid supply
			Self-supply + sell to the grid	Grid supply + self-supply backup	Mines sell collectively to the grid	Mines invest in the grid	Mines serve as anchor demand for IPP	
Description	Mine produces its own power for its own needs	Mine provides power to community through mini-grids or off-grid solutions	Mine produces its own power and sells excess power to the grid	The mine is first connected to the grid and is moving into own-generation when more economical	Coordinated investment by a group of mines, producers, and users in one large power plant off-site connected to the grid	Mine invests with government in new, or in the upgrading of, power assets under different arrangements	Mine buys power from an IPP and serves as an anchor customer	Mine does not produce any power, but buys 100% from the grid
Main generation drivers	Diesel, HFO	Diesel, HFO	Coal, Gas, Hydro	Diesel, HFO	Diesel, HFO, Solar	Hydro, Gas	Any	Any
Presence	Mali and Guinea (hydro) Sierra Leone and Liberia (oil)	Guinea, Madagascar	Zimbabwe Mozambique, Cameroon	Congo, Dem. Rep. Tanzania	Ghana	Niger Congo, Dem. Rep.	South Africa	Mozambique Zambia

Source: Africa Power–Mining Database 2014, World Bank, Washington, DC.

Note: CSR = corporate social responsibility; HFO = heavy fuel oil; IPP = independent power producer.

Grid supply is the dominant power-sourcing arrangement among mining projects, but self-supply has risen impressively, from only 6 percent of projects before 2000 to 18 percent in 2020. Among power-sourcing arrangements over 2000–20, self-supply rose the fastest at 11.5 percent, compared with 5.8 percent for intermediate options and 4.7 percent for grid supply. Even so, self-supply remains a minority among the three arrangements in 2020.

Self-supply, intermediate options, and grid supply have all registered rising total consumption to 2020, underpinned by a growing number of projects (figure O.4a). Total consumption will remain highest for grid supply to 2020. Except for intermediate options, the trend since 2000 is for average power consumption to rise by 2 percent (figure O.4b). Accordingly, while total consumption will rise for self-supply, it will fall dramatically per project, suggesting that only projects with smaller power requirements will rely on it.

Complex Factors Determine Mines' Power-Sourcing Arrangements

Three main factors determine power-sourcing arrangements. First, a country's primary energy source is important for adequate and reliable supply. Hydropower-based grids will favor integration, with mines either connected to the grid or in an arrangement with the utilities. Gas-based grids tend to have the effect of hydropower-based grids but are still rare. Two massive coal mines in Mozambique, Vale's Moatize and Rio Tinto's Benga, are planning to sell electricity from their coal-fired power plants back to the grid. Second, the supply–demand imbalances and resulting power shortages push some arrangements. For instance, the option of connecting mines to the grid and moving into

Figure O.4 Range of Power-Sourcing Options for Mines

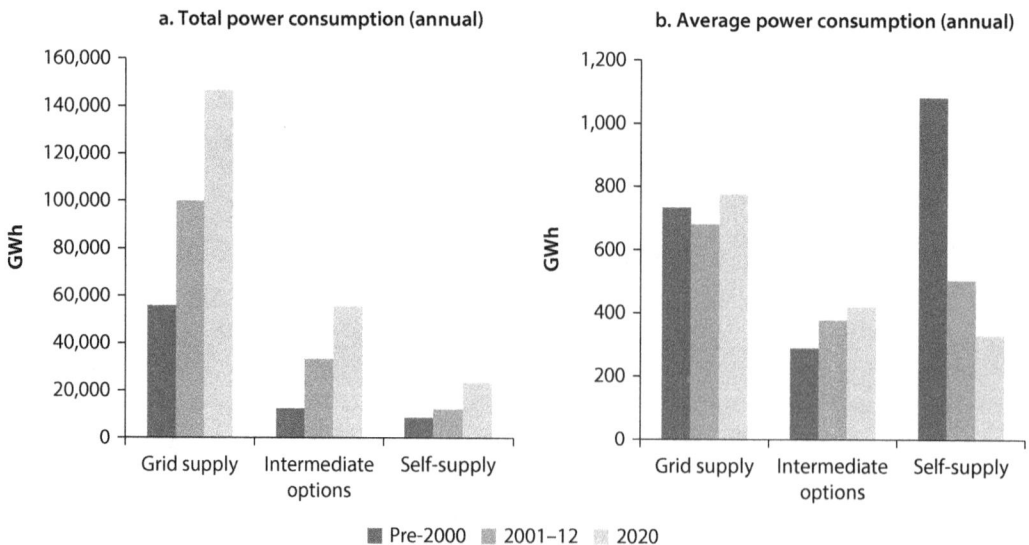

Source: Africa Power–Mining Database 2014, World Bank, Washington, DC.
Note: Average power consumption of mines is calculated as the total consumption divided by number of projects.

self-supply when reliability is poor has been exercised in more projects in South Africa and recently in Tanzania. Third, cost differentials between grid tariffs and self-supply costs play a substantial role. Self-supply occurs in oil-fuel-based grids (particularly in Liberia and Sierra Leone) where mines self-supply at a lower cost than the grid tariff—but also in hydropower-based countries where mines self-supply at a higher cost than the utilities because of frequent outages. Because of the cost of outages, self-supply remains cost-effective. Historically, mines invest with government in existing or new power assets mainly in countries with hydro-based grids or low-cost electricity. Their expectation is to build a stronger grid to deliver stable and cheap electricity in the medium to long run.

Discussions with mining companies reveal a complex set of objectives and motivations in considering power options. Most important, mines are at least as concerned about security of supply as they are about cost. They invest in self-supply even when the cost per kilowatt delivered is much higher to ensure control and continuous power availability. For example, public hydropower stations with already written-down capital costs can provide mines low-cost electricity, but if such supplies are unreliable, self-generation becomes an option.

Self-Supply Is a Loss to Economy, Utilities, and Mines

Over 2000–12, mining companies invested about $1.3 billion in generation capacity for 1,590 MW in arrangements with some form of self-supply (self-supply, self-supply and CSR, self-supply and sell to the grid, grid supply and self-supply backup). It is expected that 10,260 MW will be added to meet mining demand to 2020. Of this amount, self-supply arrangements will contribute 1,753 MW in the high-probability scenario and 3,061 MW in the low-probability scenario. This will cost between $1.4 billion and $3.3 billion and constitute 21–30 percent of generation capacity to meet mining demand in 2020. Most of this generation capacity will be thermal (diesel and HFO), while the remaining 6,650–7,195 MW is expected to be met by collective arrangements (mines sell to the grid, mines invest in the grid, and mines serve as anchor demand for IPP) and grid supply. Driving this capacity will be investments in hydro and gas-fired power generation.

These self-supply investments benefit the mines, albeit at a high cost. Moving away from self-supply can benefit utilities and the population and create new opportunities for the private sector.

Power–Mining Integration Can Reduce Costs, Benefit Communities, and Encourage Private-Sector Participation

Harnessing economies of scale can produce cost savings for both the mines and local populations. In Guinea, Mauritania, and Tanzania, there is substantial potential for using mines as anchor consumers for local electrification. In Guinea and Mauritania, where the grid is nascent, it is not yet economical to connect mines to the grid. The mines, which are contiguous and propose using self-supply, could join together or contract with an IPP to manage the generation and transmission system in a mini-grid operating at high voltage. This could occur through hydropower projects, as in Guinea, or through CCGT projects, as in Mauritania.

Three scenarios are simulated: self-supply based on diesel generation (scenario 1), shared supply among the mines (scenario 2), and shared supply for mines and neighboring communities (scenario 3). For Guinea, the capital cost (including generation and transmission) is estimated at about $310 million for scenario 2 and $592 million for scenario 3. For Mauritania, the total investment is about $104 million for scenario 2 and $142 million for scenario 3.

The cost of providing electricity through a shared plant to meet mining and local demand is much lower, offering substantial cost savings. In Guinea, the levelized cost of scenario 2, at 4.9 cents per kilowatt-hour (kWh), is the lowest, although that of scenario 3 is not much higher, at 5.0 cents per kWh (figure O.5a). In Mauritania, the cost of scenario 3, at 9.2 cents per kWh, is similar to that of scenario 2 (figure O.5b). Mines in Guinea can benefit from $640 million in cost savings and those in Mauritania from $990 million in operating costs. In scenario 3, 5 percent of Guinea's population (about 540,000 people) and 4 percent of Mauritania's population (about 143,600 people) would benefit from electrification.

For residents of the newly electrified towns, the economic benefits per household could total about $433 for Guinea and $285 for Mauritania over the project life, based on the avoided cost of dry-cell batteries.

Figure O.5 Levelized Cost Estimates of Integration Scenarios

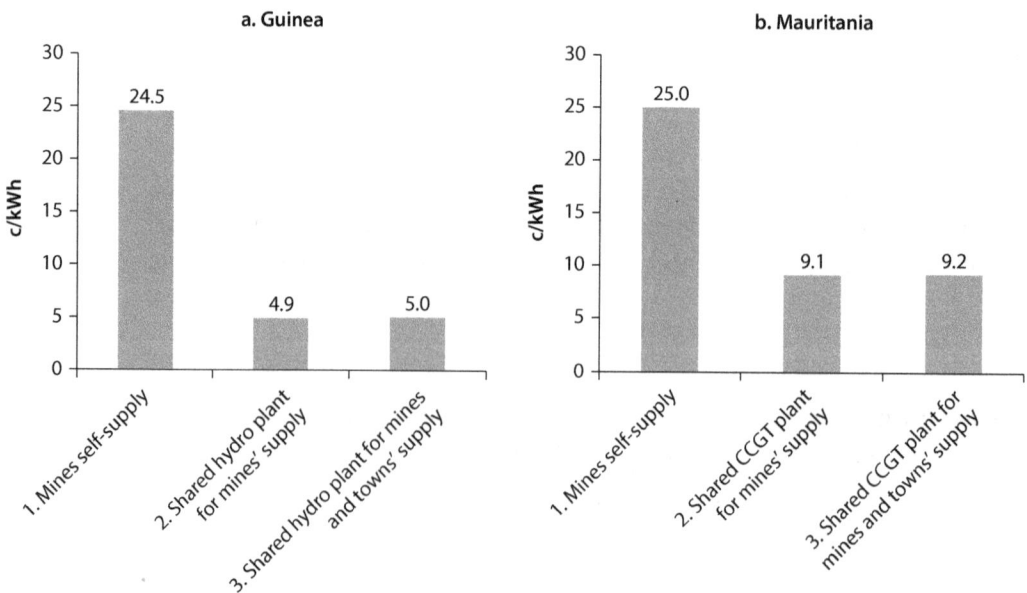

Source: World Bank.
Note: CCGT = combined-cycle gas turbine.

In Tanzania, the Tanzania Electric Supply Company's performance and financial situation are forcing many mines to move into a self-supply mode. However, mines could represent the viable off-taker needed to invest in large power generation. They could supply communities near the mines or close to the mines' grid connection that would otherwise not receive power. Integration could also involve one of three generation scenarios: a 300-MW hydropower plant at Rufiji River Basin, a 300-MW gas-powered plant in Dar es Salaam, or a 300-MW coal-fired power plant near the coal mines in Mbeya. This capacity is more than the participating mines need—the excess capacity would feed the national grid to meet rising power demand in Mwanza and Shinyanga, spreading the costs over a larger demand and resulting in the lowest unit costs for all three options for scenarios 2 and 3 (figure O.6). The cost savings for the mines would total about $3.4 billion to $3.7 billion, and consumers in Mwanza and Shinyanga would benefit from better electricity supply.

Power–Mining Integration Can Add Momentum to Regional Power Integration

Mines in Cameroon and Mozambique can provide the anchor demand to develop regional generation projects. In Cameroon, the government has a framework requiring a long-term planning and investment commitment by large power users to develop the country's hydropower resources. Private developers must compete for hydro sites, except when the site may more sensibly be allocated to

Figure O.6 Levelized Cost Estimates of Integration Scenarios, Tanzania

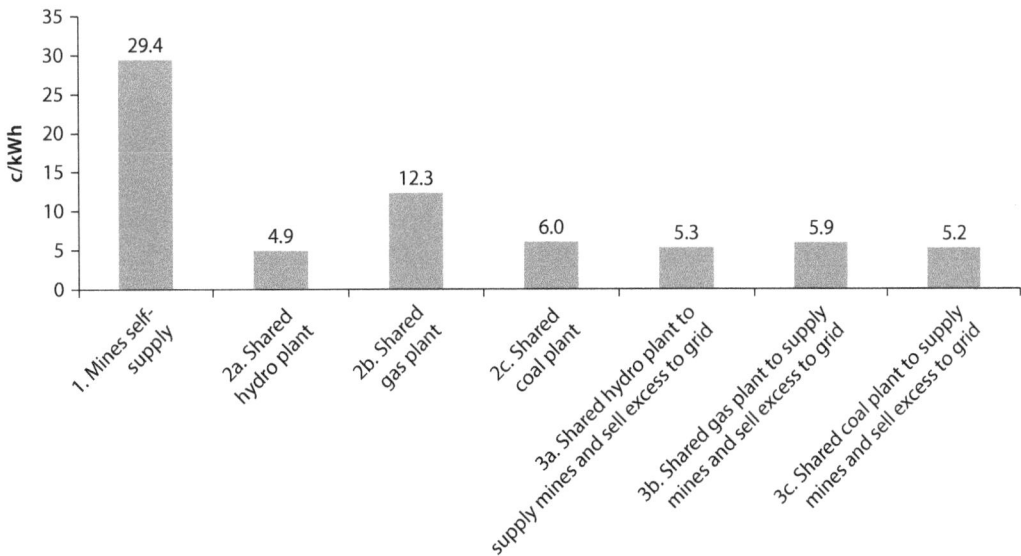

Source: World Bank.

a mine to develop power for its own needs. The challenges of ensuring that full economies of scale are being realized—and determining the optimal sharing of power between the mine and the grid—are neatly solved by requiring that the full potential of the hydropower site be developed by the mine, with the surplus sold to the grid at cost-recovery tariffs determined by the regulator. The surplus will initially be absorbed in the domestic market, but in due course, power will be available for export to the currently dormant Central African Power Pool.

In Mozambique, mines are producing high-quality coking coal for export, and the discard (waste) coal is available for power generation. Mine demand provides the rationale and anchor for developing discard coal–fired power stations, but any additional power generation capacity depends on finding purchasers in the Southern African Power Pool (SAPP). Two scenarios are simulated. In scenario 1 self-supply and discard coal–fueled power plants at each of the mines meet only mine demand (236 MW). This is merely for illustration; in practice, some export of power to the national and regional markets would be expected. In scenario 2 coal mines raise the capacity of their power plants and send their surplus through a dedicated transmission line to a 900-MW aluminum smelter at the Macuze port (1,810 MW). This scenario considers creating a market for discard coal–fired power, a smelter that could be built at the port of Macuze, which is being constructed to handle bulk coal exports from Tete. The cost of this investment, including the smelter, would total $3.5 billion. The smelter would reduce the cost to 4.1 cents per kWh, compared with 5.8 cents per kWh under scenario 1 (figure O.7). While the regional power market is the most likely option for the discard coal, this smelter could be an alternative. But carbon dioxide (CO_2) emissions are considerable, at 11,837 tons in scenario 1 and 55,071 tons in scenario 2.

Figure O.7 Levelized Cost Estimates of Integration Scenarios, Mozambique

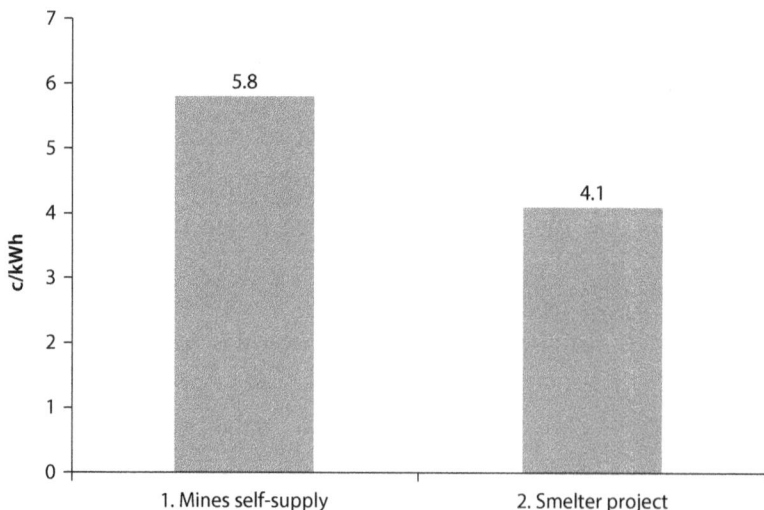

Source: World Bank.

Technical and Financial Constraints Must Be Addressed to Facilitate Power–Mining Integration

The primary physical constraint is lack of a national transmission grid capable of catering to additional flows as the mining sector and the rest of the economy expand. In Mozambique, transmission constraints are pressing for any generation option, including coal-fired plants, to provide power to Electricidade de Moçambique (EDM) and access the regional SAPP market. Attaining full potential in the Democratic Republic of Congo and Zambia requires addressing the regional power market and interconnector constraints.

Another major constraint is the utilities' weak financial situation. In the Democratic Republic of Congo, revenue was collected from only about 1 kWh for each 2 kWh produced in 2012; coupled with low tariffs, electricity sector inefficiencies were estimated at up to 4 percent of GDP. In Tanzania, the Tanzania Electric Supply Company (TANESCO) has run financial deficits for many years; the operating loss in 2012 was $139 million, bringing cumulative losses to $503 million or 2 percent of GDP.

There are also country-specific constraints. In Ghana, the desire to diversify from hydropower through gas generation is limited by unreliable gas supplies. Cameroon has been constrained by the control of flows on the key hydropower river. This problem is expected to be resolved soon, so the key issue is strengthening the framework—established to foster mining companies' expected investments in large hydropower projects—and making it operational. In Guinea and Mozambique, issues involving transport, more than electricity capacity, constrain the expansion of iron ore mining and coking coal exports. But expanding rail and port capacity would help integrate the power and mining sectors.

Lessons of Experience and Risks of Engagement Must Be Carefully Considered

The potential for power–mining integration is substantial, but there is little indication that such integration has benefited local populations in the past. In mineral-rich countries such as the Democratic Republic of Congo, Guinea, Mauritania, Mozambique, Tanzania, and Zambia, electrification rates remain below 20 percent. There are a few instances of mines participating in electrification through CSR. But mines invest in the power network through a range of commercial arrangements. In such cases, the infrastructure often belongs to the national utility and a prepayment is treated as a loan, which is repaid in kind rather than in cash through an offset in invoicing for power purchased. This could be made equivalent to a discounted tariff during the repayment period. But many of these arrangements remain unknown to the public and media, leading to transparency concerns and damaging public perception of mining companies.

Tension from tariffs has been undermining the power–mining relationship. In Cameroon, the Democratic Republic of Congo, and Zambia, the grid tariff is less than 10 percent of self-supply with a typical 5-MW diesel generator.

Zambia has by far the lowest mining tariff, a set of negotiated prices among the Zambia Electricity Supply Corporation, the Copperbelt Energy Corporation, and individual mines. Mining tariffs are lower than long-run marginal cost, but efforts are under way to bring mining tariffs to cost-reflective levels by 2014. Similarly in Ghana, mines have not had a financial incentive to sign power purchase agreements (PPAs) with IPPs because they receive low-cost hydro or blended hydro from the generator, Volta River Authority (VRA) (World Bank 2013b). Recently, however, tariffs have risen and are higher than long-run marginal cost.

As a partner to power sector development, mining involves some risks, which could explain why power–mining integration has been limited to date. First, planned investments in mining may not materialize because of price swings, difficulties in raising capital, overly optimistic geological assessments, and political instability. Prices in international commodity markets fluctuate, sometimes wildly. The period since 2003 has seen the biggest sustained upswing, though prices have moderated since 2012 (figure O.8). Second, when prices fall, mines and smelters may cut their output and thus their power needs. Third, because mines have finite lives, usually shorter than those of large power facilities, power investments will eventually need other customers, who may not materialize.

Figure O.8 Mineral Prices, 1983–2013

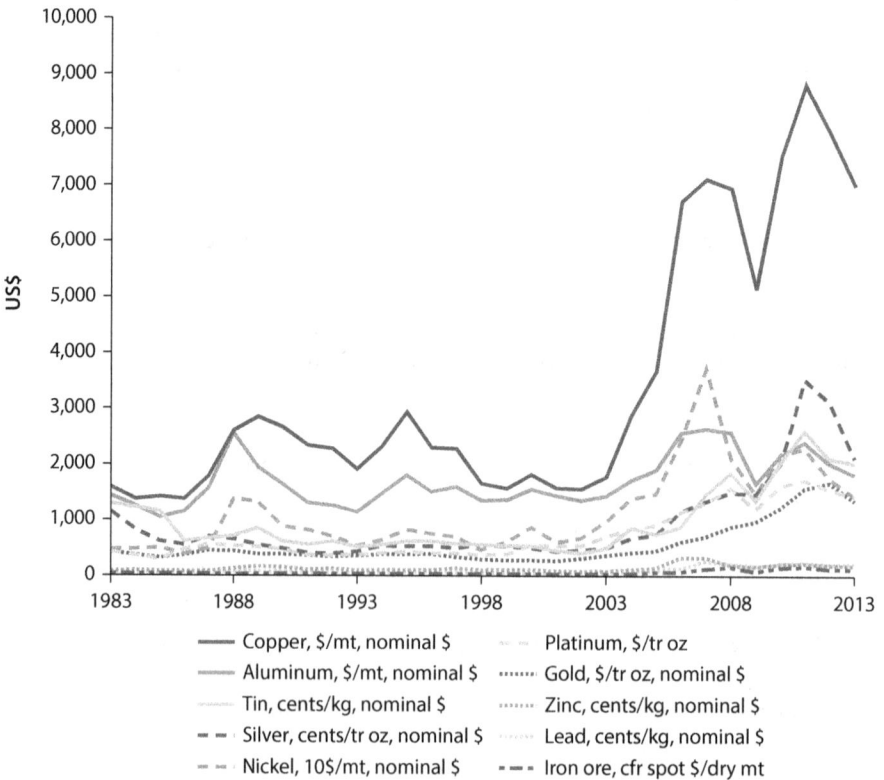

Source: World Bank 2013a.

Fourth, mining can become a powerful lobby to extract subsidies or special privileges from the power sector, particularly if electricity demand grows and mining operations are no longer needed as anchor customers. If this occurs, mining demand may crowd out medium-size firms and residential consumers, reducing the possibility for extending electricity access.

Many institutional roadblocks stymie power–mining integration. Resource pooling among mines is difficult given mining's highly competitive environment. There is little motivation to construct power plants with greater capacity than a mine's own demand without regulatory and commercial incentives and a transmission network with spare capacity. Supplying power to local communities is usually overlooked unless mining companies integrate this in their CSR work or are contractually obligated to do so. And many public utilities are not viable partners for mining companies to invest in.

The Government and Policymakers Must Seize This Opportunity and Adopt Appropriate Risk-Mitigation Mechanisms to Create a Win-Win Situation for All Parties

Strengthen power sector finances. Public utilities need to be perceived as viable and creditworthy partners of the mining companies. Regardless of the power-sourcing arrangement, the utility will be the mines' main partner, so ensuring its financial health and creditworthiness is essential to successful and sustainable integration. Whether the public utility acts as a power off-taker, power distributor, or co-investor, its financial health will determine the range of possible power-sourcing arrangements.

Support the operating environment for IPPs. National power sectors must be sufficiently liberalized to allow for IPPs in the generation segment and preferably also encourage the private sector to invest in transmission. The mines are big players in all case study countries: they have the capacity to invest directly in power generation and transmission or to do so as the anchor customers for private IPPs. The framework for IPP generation is already in place in all countries except the Democratic Republic of Congo, where new legislation is in the final stages of presidential endorsement.

Integrate mining demand in power sector planning. Current and projected mining power demand should be included in the supply–demand analysis and traditional least-cost power plans once concrete agreements with mines are in place. To date, only Tanzania has explicitly incorporated mining growth and its power investment plans in its power master plan. Among the power pool master plans, only the West Africa Power Pool has distinguished mining demand. This has led to situations where the mining investment cannot be leveraged because the transmission network is not adapted to carry the load that mines could sell back to the grid. An integrated power sector plan would allow for a much clearer vision of a country's future power situation. To facilitate this integration, a country's mining ministry (or relevant department if mining and energy are in the same ministry, as in Cameroon, Mauritania, and Tanzania) should be included.

Power requirements should be integrated into mining law. New mining operations should be required to outline their power demand and how they will source it. When collaboration is possible, the law should require dialogue between the mining company and the relevant government agencies. The focus should be on dialogue, not mandated actions. But for mines of a certain size that self-supply in remote areas, a requirement to supply local communities should be considered.

Source expertise. Governments should take a long-term perspective when identifying potential synergies and create an attractive enabling environment. Short-term solutions will often be incompatible with long-term country needs, particularly when rising demand from nonmining sources is taken into account. Many institutional arrangements are possible, and a one-size-fits-all approach should be avoided. Governments should seek expert advice to help identify the arrangements that will bring their countries the greatest benefits.

Strengthen regulatory mechanisms. Countries need to move toward a more effective regulatory structure. In most countries, technical regulation is satisfactory, but economic regulation appears weak in setting cost-recovery tariffs that enable the utilities to maintain equipment and make investments. In addition to economic regulation, an effective regulatory system must manage risks and regulate access. Risks in IPPs and PPAs include defaults and delays in payments and failing to honor contract obligations. Effective regulators enforce contracts and strengthen the utilities that cannot provide sovereign guarantees (for example, escrow accounts, profit repatriation, and guarantees against nationalization), which are usually required when the utility is not viable.

Regularly review mining tariffs. Mine tariffs should be carefully set and regularly evaluated. Using large mining operations as anchor customers for large power development is attractive but requires caution. Flexible power pricing can prevent mines being subsidized at the expense of the utility—or taxpayers. Nonmining industries and residential consumers will eventually want part of the power consumed by the mining company. Mining's large power demand will crowd out other consumers, even if they are willing to pay a higher price for it. Contracts with the anchor customer must consider this possibility.

Carefully draft CSR contracts. Some mining companies include local electricity provision in their CSR policy. But the voluntary nature of this activity can undermine sustainability and government responsibility. If mines and governments move ahead with the CSR arrangement, developing model concession agreements that mandate providing electricity within a certain radius would raise investor certainty, put all mining companies on an equal footing in their CSR programs, and enhance the accountability of government as the contract enforcement authority. An important part of such arrangements is building capacity to maintain the local distribution system, which allows the government or its contractors to take charge of distribution and bill collection and ensure future sustainability.

Use regional platforms. In the short and medium term, a country's own mining and nonmining power demand may not justify what could otherwise be seen as an optimal power investment. Full benefit from the new arrangements will require regional coordination on infrastructure and power-sharing policies.

The World Bank Group Must Support Governments' Efforts to Harness the Synergies Offered by Mining Power Demand

The World Bank Group (WBG) can engage directly in project ideas explored as scenarios in this study, particularly for countries with inadequate and unreliable grids. These projects can be developed with mines as anchor consumers, but they will need government and donor support because mines have little incentive to create oversized generation plants to serve neighboring populations or send excess capacity to the grid. The WBG could also provide technical inputs and innovative financial instruments (such as guarantees and credit enhancements) to facilitate commercial arrangements among mines and between mines and utilities. Where mines are largely grid-supported, the WBG can offer analytic studies and timely advice on rationalizing tariff structures and negotiating long-term tariffs.

The WBG should continue supporting the power utilities and regulatory agencies to improve not only the power sector's financial health but also prospects for the private investment required to meet the dramatic growth in demand. WBG staff should support regular collaboration between the power and mining units, including joint missions in selected countries. They should also promote interaction between the country's power and mining stakeholders through national and local workshops and conferences to ensure that the various groups are aware of the possibilities, challenges, and responsibilities. In addition, they should promote regional dialogue on harnessing large energy resources using mining companies as anchor consumers and supporting regional interconnection projects. In some subregions this could be accomplished through regional power pool platforms.

Revenues generated by the mining sector via taxes, dividends, and/or royalties offer the potential to serve as an important driver of infrastructure development, if properly managed. Financial tools exist to facilitate delivery of public infrastructure services. Securitization can serve to improve borrowing conditions to investment grade and reduce financing cost; mitigate revenue uncertainty via risk sharing; promote strong governance, transparency, and solid asset management practices; and develop a capital market track record benefiting future transactions. A toolkit could be developed to provide in-depth analysis of the institutional arrangements presented in this study, including methodologies for evaluating costs and benefits. The wealth of information in the Africa Power–Mining Database will need regular updating. The WBG should identify an African institution to host and maintain this database in a publicly available space.

Finally, a technical advisory facility could be considered to help governments negotiate contracts between mining companies and power providers. Given the strong disadvantage of most African governments, this facility's technical inputs could contribute to leveling the playing field.

References

Eberhard, Anton, Orvika Rosnes, Maria Shkaratan, and Haakon Vennemo. 2011. *Africa's Power Infrastructure: Investment, Integration, Efficiency*. Directions in Development. Washington, DC: World Bank.

Kumar, Arun, Tormod Schei, Alfred Ahenkorah, Rodolfo Caceres Rodriguez, Jean-Michel Devernay, Marcos Freitas, Douglas Hall, Ånund Killingtveit, and Zhiyu Liu. 2011. "Hydropower." In *IPCC Special Report on Renewable Energy Sources and Climate Change Mitigation*, edited by Ottmar Edenhofer, Ramon Pichs-Madruga, Youba Sokona, Kristin Seyboth, Patrick Matschoss, Susanne Kadner, Timm Zwickel, Patrick Eickemeier, Gerritt Hansen, Steffen Schlömer, and Christoph von Stechow, 437–96. Cambridge, U.K.: Cambridge University Press.

Santley, David, Robert Schlotterer, and Anton Eberhard. 2013. *Harnessing African Gas for African Power*. Washington, DC: World Bank.

Wilburn, D. R., and K. A. Stanley. 2013. "Exploration Review." United States Geological Survey, Reston, VA.

World Bank. 2012. *Africa—Second Additional Financing for Southern African Power Market Project: Restructuring*. Washington, DC. http://documents.worldbank.org/curated/en/2012/06/16360805/africa-second-additional-financing-southern-african-power-market-project-restructuring.

———. 2013a. *Enterprise Surveys*. Washington, DC. http://www.enterprisesurveys.org/.

———. 2013b. "Energizing Economic Growth in Ghana: Making the Power and Petroleum Sectors Rise to the Challenge." Washington, DC. http://issuu.com/fosu/docs/energizing_economic_growth_in_ghana.

CHAPTER 1

A High-Risk–High-Return Opportunity

The Anchor Consumer's Role in Developing the Power Sector

Half of the world's power access deficit resides in Sub-Saharan Africa (SSA). Lack of access to modern energy services poses major challenges for maintaining the economic growth needed to increase equality and reduce poverty. The region's overall electrification rate is only 31 percent, varying from 60 percent in urban areas to just 14 percent in rural areas. To meet their basic lighting needs, households without electricity access must depend on kerosene and dry-cell batteries, which have detrimental health and environmental impacts. Even households with electricity access consume very little. The average annual per capita energy consumption in SSA (excluding South Africa) is 155 kilowatt-hours (kWh), compared with 4,470 kWh in South Africa and 13,000 kWh in the United States.[1]

In 2011 the region's installed generating capacity was only 78 gigawatts (GW), the lowest among all regions. South Africa accounted for 45 GW, distantly followed by Nigeria at 6 GW. The generating capacity for the rest of SSA was 33 GW, equivalent to that of Norway or Argentina. At current investment levels, the number of Africans without electricity will rise from 590 million in 2013 to 655 million in 2030. Power demand is expected to grow 6 percent a year, so massive investments—many times higher than the business-as-usual scenario of little more than 1 GW a year—will be needed to keep pace with the growing continent's aspirations. Increasing the region's per capita energy consumption to the current consumption level of South Africa by 2030 would require about 1,000 GW of new generation capacity (Bazilian and others 2012). Addressing Africa's chronic power problems and lack of electricity access will require major investments in expanding and refurbishing power infrastructure.

SSA's households and small businesses with electricity access pay a heavy price for the region's inadequate and unreliable power supplies. They pay three times as much as their wealthier counterparts in the United States and Europe, and regularly experience power outages (Foster and Briceño-Garmendia 2010). Nearly half of SSA firms—surpassed only by those in South Asia and the

Middle East and North Africa—identify electricity as a major constraint to doing business (figure 1.1a). Pervasive outages cost about 5 percent of annual sales (figure 1.1b), typically representing 1–4 percent of a country's gross domestic product (GDP) (Foster and Briceño-Garmendia 2010). About 44 percent of SSA firms—10 percent fewer than in the Middle East and North Africa—cope by owning or sharing a generator, fueled typically by diesel or heavy fuel oil (HFO) (figure 1.1c). It takes 141 days for SSA businesses to receive a new electricity

Figure 1.1 Unreliability of Power Supply and Coping Mechanisms

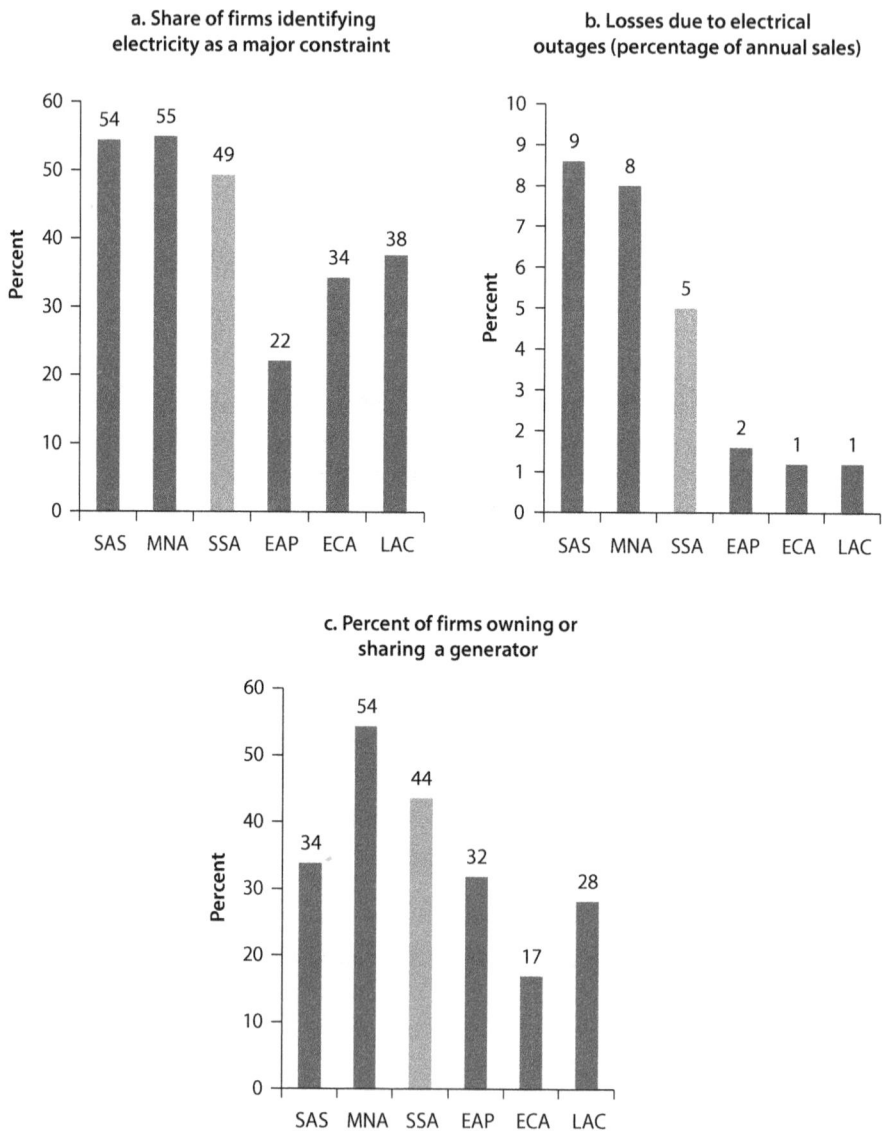

a. Share of firms identifying electricity as a major constraint

b. Losses due to electrical outages (percentage of annual sales)

c. Percent of firms owning or sharing a generator

Source: World Bank 2013.
Note: EAP = East Asia and Pacific; ECA = Eastern Europe and Central Asia; LAC = Latin America and the Caribbean; MNA = Middle East and North Africa; SAS = South Asia; SSA = Sub-Saharan Africa.

connection—only marginally less time than in South Asia (145 days) and Eastern Europe and Central Asia (146 days)—and much longer than in advanced economies (World Bank 2013).

This suboptimal situation coexists with SSA's vast energy resources. Lack of financing, low power consumption per capita, and the small scale of most national power systems have prevented utilities and private companies from harnessing these resources. Hydropower is Africa's most abundantly available energy resource, followed by coal and natural gas (map A.1). In Cameroon, the Democratic Republic of Congo, Ethiopia, Guinea, and Zambia, hydro can produce large-scale power at less than one-third of the cost of thermal generation, but only 8 percent of its potential has been developed, compared with 18 percent in Latin America and the Caribbean and 20 percent in Asia (Kumar and others 2011). Gas—found along the East Africa coast (Mozambique and Tanzania) and in some West African countries (Angola, Côte d'Ivoire, and Nigeria)—together with the geothermal capacity of East Africa's Rift Valley, can provide additional power generation.

When the top six SSA countries with the greatest potential in each of six resources (hydro, gas P1, gas P2[2], coal, geothermal, and oil) are ranked, two attributes stand out (table 1.1). First, the country with the most abundant supply of each energy resource varies, though some countries have abundant supplies of more than one resource. Angola and Nigeria are among the top performers in four categories and have the most abundant energy sources. Equatorial Guinea, Mozambique, and the Republic of Congo perform well in three categories. Second, energy resources are heavily concentrated; for each resource, the top six countries have a combined share of SSA energy resources ranging from 76 percent (hydro) to 100 percent (geothermal).

The role of anchor consumers is vital to aggregating the demand for resources that will require large-scale generation projects. SSA's energy resources are concentrated in countries that would not find it commercially viable to limit themselves to meeting household and industrial demand. Most new power generation capacity will be driven by commercial and industrial customers, characterized by predictable load and requiring continuous power delivery and large demand for electricity. On average, commercial and industrial customers represent half of all sales revenue (Foster and Briceño-Garmendia 2010). These anchor customers could help fill the power sector investment gap as viable off-takers with large energy demand that could reduce the investments' financial risk. More important, anchor loads could attract the private sector. Private sector participation in SSA, unlike in other regions, has been limited. To date, about 23 medium- to large-scale projects have been initiated, contributing about 4.1 GW of capacity in 11 countries (ICA 2011). This share is miniscule compared with existing or required capacity. Given the large scale of power needs, it is necessary to create an attractive private sector operating environment in which anchor consumers can play an important role.

Leveraging mining power demand and its capital investments in power infrastructure can help develop the power sector of Africa's mineral-rich economies.

Table 1.1 Energy Resources in Sub-Saharan Africa

	Hydro		Natural gas				Coal		Geothermal		Oil	
	Technically feasible		*P1 reserves (2013)*		*P2 reserves (2013)*		*Hard coal and lignite, remaining potential*		*Geothermal potential*		*Proved reserves (2012)*	
		GW		*Bcf*		*Bcf*		*Million tons*		*MW*		*Million barrels*
	Congo, Dem. Rep.	101	Nigeria	110,300	Nigeria	132,360	South Africa	33,896	Kenya	10,000	Nigeria	37,200
	Ethiopia	63	Angola	12,000	Angola	18,000	Zimbabwe	25,502	Ethiopia	5,000	Angola	12,667
	Madagascar	44	Congo, Rep.	3,200	Mozambique	4,490	Mozambique	24,187	Rwanda	700	Republic of South Sudan	3,500
	Angola	36	Mozambique	2,700	Equatorial Guinea	4,000	Botswana	21,240	Uganda	450	Gabon	2,000
	Cameroon	28	Equatorial Guinea	2,649	Congo, Rep.	3,200	Swaziland	4,644	Tanzania	380	Equatorial Guinea	1,705
	Gabon	19	Tanzania	1,369	Tanzania	1,709	Nigeria	2,740	Burundi	300	Congo, Rep.	1,600
Total SSA		383		136,146		168,657		118,639		16,830		63,673
Top 6 as % of SSA		76		97		97		95		100		92

Sources: UHD 2009; IFC 2011; Andruleit and others 2012; ECA 2013.

Note: P1 = proven reserves (developed and undeveloped); P2 = P1 and probable reserves; SSA = Sub-Saharan Africa.

Mining can be an anchor consumer, given its status as one of SSA's leading indus-trial activities, contributing substantially to exports, fiscal revenues, and growth. There is precedence for using mining as an anchor consumer. For example, in Ghana in 1966–96, the Volta Aluminum Company Limited (VALCO) smelter was the anchor customer for the Akosombo Dam, the Volta River Authority's centerpiece investment; the dam supplied most of the country's electricity needs, as well as a portion of those for neighboring countries. But by the late 1990s con-sumer and other industrial demand had grown so much that the smelter became a drag on the system and had to close. Similarly, in Zambia, much of the country's power development originally used mining operations as anchor customers, but two new hydropower projects in the country are financed partly by a 30 percent rise in power tariffs for mines in the Copperbelt.

Mining's Contribution to Socioeconomic Development

The SSA region hosts substantial amounts of the world's mineral reserves in value terms, including 22 percent of the gold, 58 percent of the cobalt, 95 per-cent of the platinum group metal (PGM) reserves, 7 percent of the copper, and 18 percent of the uranium (USGS 2013). Recent discoveries in this century have also turned SSA into a potentially important region for iron ore and coal. The region already depends heavily on mining; in 2010 mineral products accounted for more than 20 percent of exports for 18 countries and more than 40 percent of exports for 13 countries (figure 1.2). Fiscal revenues from mining accounted for more than 20 percent of the total for Botswana, the Democratic Republic

Figure 1.2 Mineral Dependence in Africa, 2010

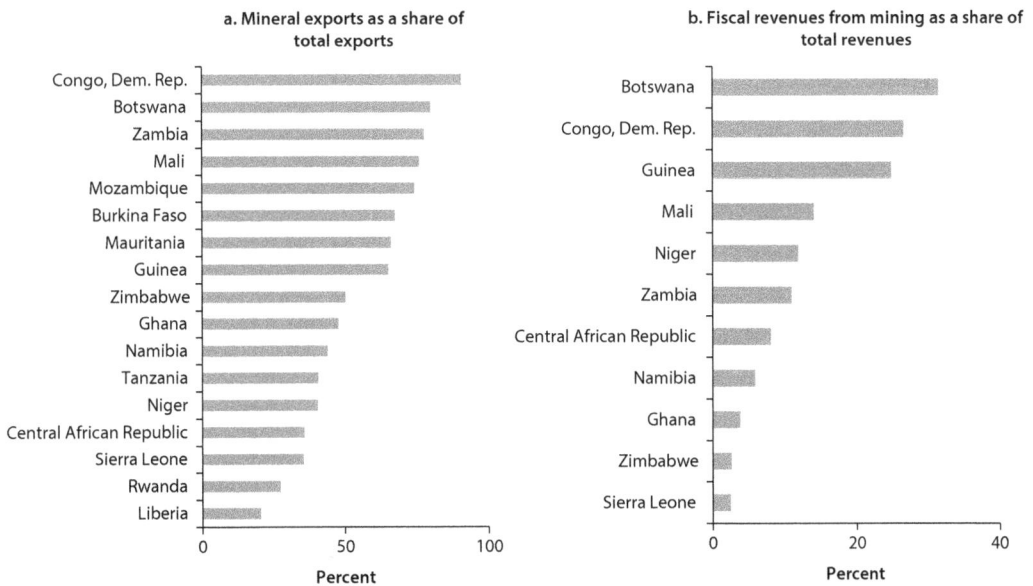

Source: IMF 2013.

of Congo, and Guinea. Yet for many of these countries, the mining sector is just beginning to develop on a large scale—numerous resources are scheduled to go into production in the next 10 years.

These impressive numbers may be only the start of great economic growth. For Guinea and Liberia, projected mining investment to 2020 can constitute an overwhelming proportion of GDP (2012) (figure 1.3). From a mineral standpoint, SSA is largely unexplored. Per unit of land, known mineral reserves are only one-fifth the average for Organisation for Economic Co-operation and Development (OECD) countries (Collier 2010). While Africa contains 20 percent of the Earth's land mass, it provides only about 5 percent of its minerals in value terms, and far less if South Africa is excluded. One main reason is that until this decade, SSA severely lagged behind the rest of the world in exploration spending. But in 2002–12, Africa's share of global mining exploration—itself steeply rising—grew from 10 percent to 17 percent. Africa's absolute amount of spending on exploration rose by more than 700 percent, reaching $3.1 billion in 2012 (Wilburn and Stanley 2013) (figure 1.4).

In the long run, a country or continent's mining potential depends on the quality of its geological resources (partly unknown), infrastructure, and amount

Figure 1.3 Mining Investment, Current and Forecast, 2000–20

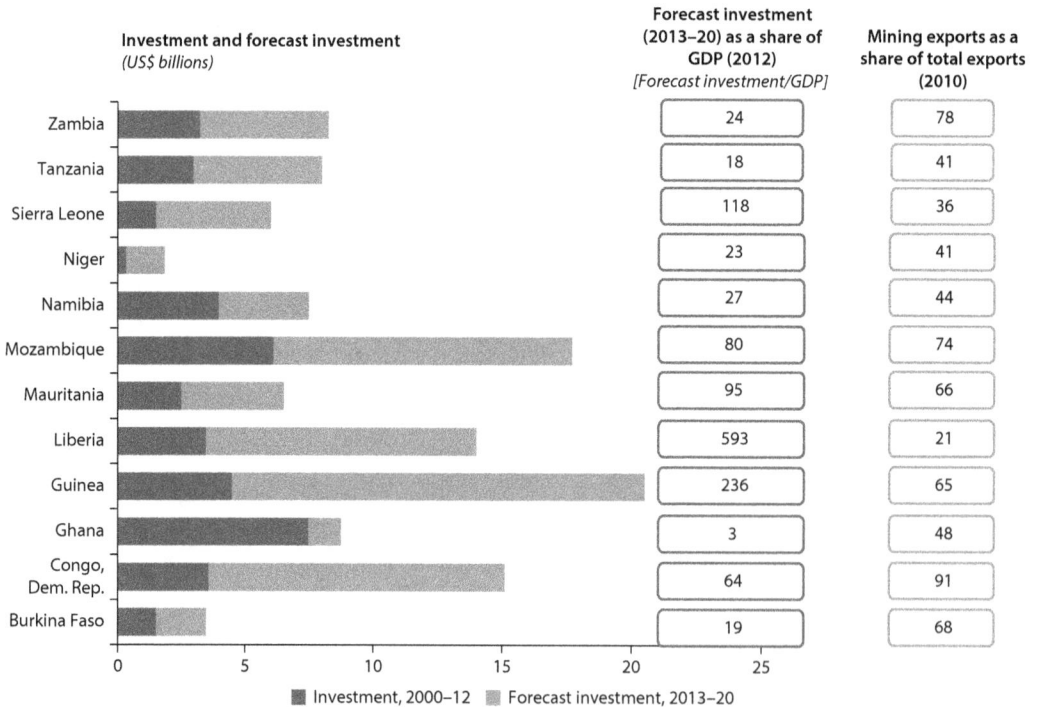

	Investment and forecast investment (US$ billions)	Forecast investment (2013–20) as a share of GDP (2012) [Forecast investment/GDP]	Mining exports as a share of total exports (2010)
Zambia		24	78
Tanzania		18	41
Sierra Leone		118	36
Niger		23	41
Namibia		27	44
Mozambique		80	74
Mauritania		95	66
Liberia		593	21
Guinea		236	65
Ghana		3	48
Congo, Dem. Rep.		64	91
Burkina Faso		19	68

■ Investment, 2000–12   ▨ Forecast investment, 2013–20

Source: Based on World Bank 2012a.

Note: The investment figures are averages across a range; for Mozambique, forecast investment excludes natural gas.

Figure 1.4 Regional Spending on Mineral Exploration, 2002–12

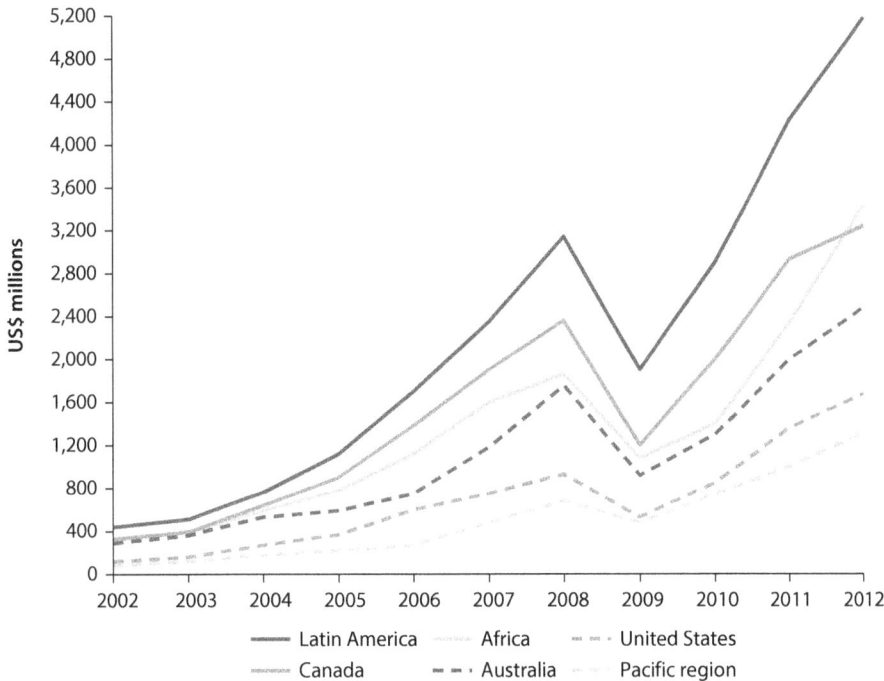

Source: SNL Metals Economics Group 2013.

of exploration. In turn, the amount of exploration depends on mining policies,[3] mineral taxation regime, access to land (legal and physical), political stability, and general security. But mineral wealth tends to be strongly related to land mass: 10 of the world's largest mining countries, by 2010 mineral value, are among the 14 geographically largest.

SSA's large land mass and rising exploration budgets strongly suggest the region's considerable potential for mining expansion, particularly north of South Africa. However, as can be seen in Guinea, Liberia, and Mozambique, mining developments are being delayed by the heavy infrastructure requirements. As this study will show, for projected mining projects to be operational by 2020, the mining sector's power requirements are expected to triple from 2000 levels. And the region's mining sector power requirements may triple again by 2040.

Many SSA countries have had rapid sustained growth since the turn of the century, due largely to the global boom in mineral prices. By the time the boom hit in 2003, Ghana, Tanzania, and Zambia, with their recently reformed mining sectors, were well positioned to take advantage of the higher mineral prices because mine development and mineral exploration had already begun to flourish. The much higher prices encouraged these trends in countries with expanding mining sectors and led other countries to undertake similar reforms, eventually leading to booming mining sectors in such countries as Burkina Faso, Mauritania, and Mozambique.

GDP growth rates of mineral-dependent countries since 1991 and projected to 2018 compare favorably with those of non-mineral-dependent countries. In 2010 mining accounted for more than 20 percent of exports in SSA's mineral-dependent countries (figure 1.5a).[4] After the 2008 global financial shock, these countries grew more than 2.5 times faster than the global average, and the current decade promises much of the same. Even though the recent decline in mineral prices has had a somewhat dampening effect on new investment, the impact on output from already operating mines has been minimal.[5] Between 2013 and 2018, GDP is projected to grow faster in mining countries than in nonmining ones.

The mining sector contributes to employment and other improvements in welfare. Although it is commonly believed that there are no jobs in formal sector mining, the reality is that mining firms buy tens to hundreds of millions of dollars of goods and services each year, the providers of which often employ several times more workers than the mines themselves. But to reap these benefits, a substantial investment in planning, policy, and the regulatory framework is needed. Well-established mining countries (such as Australia, Canada, Chile, Peru, and South Africa) have an average of 1 million mining jobs (excluding multiplier impacts). Of course, many SSA countries are relative newcomers and are only starting to build up their mining supply industries.

In the 2000s, SSA's mineral-dependent countries scored relatively well on the Human Development Index (HDI), including the disaggregated education and

Figure 1.5 Growth Rate and HDI Improvements in Mineral- and Nonmineral-Dependent Countries

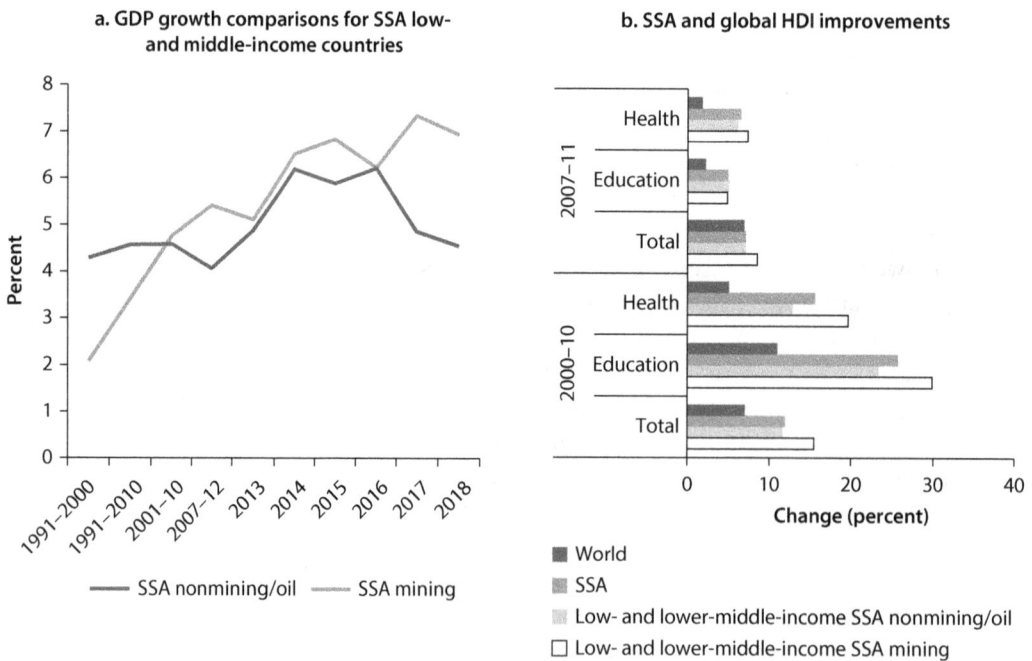

a. GDP growth comparisons for SSA low- and middle-income countries

b. SSA and global HDI improvements

— SSA nonmining/oil   — SSA mining

■ World
■ SSA
▨ Low- and lower-middle-income SSA nonmining/oil
☐ Low- and lower-middle-income SSA mining

Source: IMF 2013; UNDP 2013.
Note: HDI = human development index; SSA = Sub-Saharan Africa.

health scores. Increases in all three HDI measures were higher for the region's low- and lower-middle-income, mineral-dependent countries than for their non-mineral-dependent counterparts and the global average in all categories except education for 2007–11 (figure 1.5b).

In many SSA countries, the mining sector is a proven source of growth, as well as an important source of tax revenue that can be used to promote sustainable development. In many other countries where mining is the only large industry, it is viewed as a key engine of growth. Good mining sector management is often considered a primary path to industrialization through upstream, vertical (procurement), and downstream linkages or, in some cases, through industrial clusters and growth poles. Most mineral-dependent SSA countries now emphasize increasing mining's benefits, particularly through upstream and vertical linkages, though some are also encouraging more domestic beneficiation.[6] The work on mining or natural resource corridors stresses that in regions with several major deposits, the required infrastructure for mining operations—rail, roads, power, and ports—can be leveraged to develop other industries, including, of course, mining supply industries.[7]

Risks in Power–Mining Integration

Yet integrating power and mining is not without risks. First, planned investments in mining may not materialize because of commodity price volatility, financing challenges, faulty geological assessments, political instability, and weak contract enforcement. The period since 2003 has seen the biggest price upswing for most commodities since the turn of the last century, though prices have moderated since 2012 (figure 1.6). Price declines may cause mines and smelters to reduce their output and thus their power needs. Mines have finite lives, usually shorter than those of large power facilities, so power investments often need to rely on customers, who may not materialize. The mining sector may represent (or become) a powerful lobby that can extract subsidies or special privileges from the power sector, particularly if demand for electricity grows and mining operations are no longer needed as anchor customers. If that happens, mining demand may crowd out medium-size firms and residential consumers, in turn reducing the possibilities for extending access to electricity.

So far, this opportunity has hardly been harnessed. Among mining-dominated countries, the electrification rate, as an indicator of power-sector outcomes, remains low (figure 1.7). For example, in the Central African Republic, Liberia, and Niger, electrification reaches less than 10 percent of the population, compared with more than 80 percent coverage in Gabon and South Africa.[8] The correlation coefficient between the electrification rate and mining's contribution to GDP is relatively low, at 0.34. In such countries as the Democratic Republic of Congo, Guinea, and Mauritania, where mining's contribution to GDP is higher, the electrification rate is comparatively lower. The large mining investments planned for the Republic of Congo, Guinea, and Mauritania, combined with the ongoing challenge of electrification, suggest the need for dialogue between the power and mining sectors to find collaborative solutions.

The Power of the Mine • http://dx.doi.org/10.1596/978-1-4648-0292-8

Figure 1.6 Mineral Prices, 1983–2013

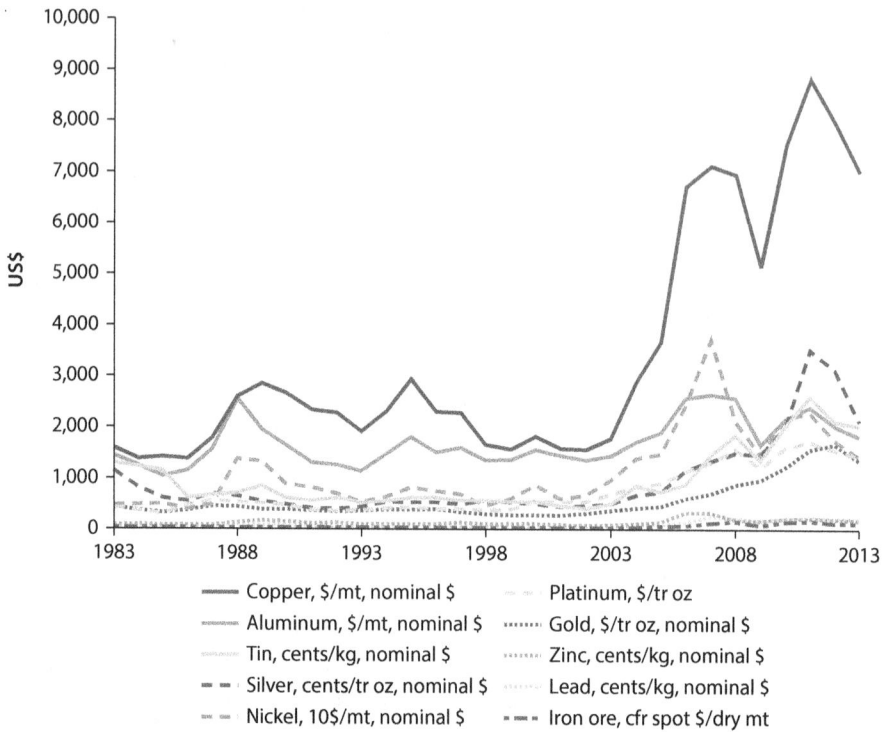

Legend:
- Copper, $/mt, nominal $
- Aluminum, $/mt, nominal $
- Tin, cents/kg, nominal $
- Silver, cents/tr oz, nominal $
- Nickel, 10$/mt, nominal $
- Platinum, $/tr oz
- Gold, $/tr oz, nominal $
- Zinc, cents/kg, nominal $
- Lead, cents/kg, nominal $
- Iron ore, cfr spot $/dry mt

Source: World Bank 2013.

Figure 1.7 Electrification Rate and Mining Contribution to GDP

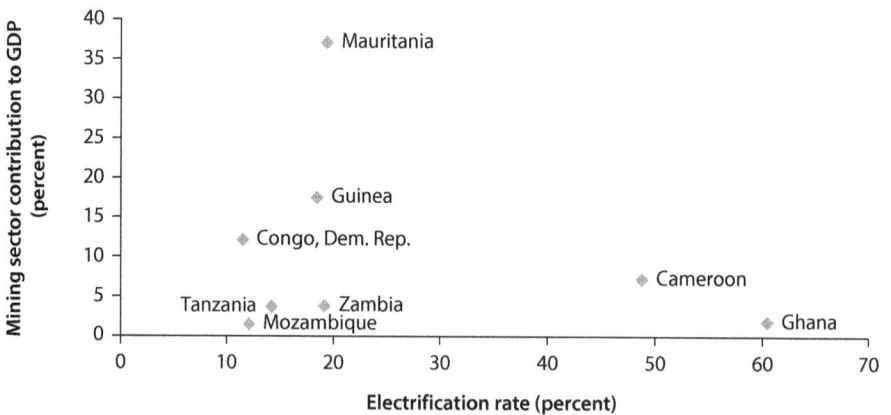

Sources: Banerjee and others 2013; IMF 2013.

Given the large projected rise in mining operations' power needs for the rest of this decade and beyond, now is the right time to take advantage of opportunities for power–mining integration. Solutions leading to inexpensive, reliable power will likely increase the number of feasible mining operations or at least the amount of local beneficiation profitable for the host country. Cheaper, more accessible power will also boost domestic mining supply companies' ability to compete with international firms. The alternative is an increasing reliance on expensive, inefficient self-supply using diesel or HFO, which, in turn, would see fewer mining operations and less beneficiation. It could also mean that other consumers, including companies competing to supply the mining industry, would find it harder to access grid electricity. New modalities of mining power generation are opening up so that the benefits of mining sector growth can be fully realized and distributed more broadly.

Scope of This Report

This study is the first to systematically analyze the potential and challenges of power–mining integration in SSA. Given the scale of the opportunity, it is essential to establish mining power demand, map the range of SSA power–mining relationships, and quantify the economic gains from harnessing these relationships. This study estimates the cost savings of various integration options and identifies the challenges and enabling conditions to realizing savings for the economy.

The study had two tracks. First, a landscape analysis was carried out across Africa to establish mining demand for power since 2000 and project demand for projects that may become operational by 2020. For this purpose, a database, known as the Africa Power–Mining Database 2014, was constructed. The database contains 455 mining projects in 28 SSA countries with ore reserve value assessed at over $250 million for each project. In addition to details on the mining property, type of mineral, and stage of processing, the database identifies the typologies of power-sourcing arrangements for these projects. Second, a case study analysis of various options for power–mining integration was conducted for eight mineral-rich economies: Cameroon, the Democratic Republic of Congo, Ghana, Guinea, Mauritania, Mozambique, Tanzania, and Zambia. For the group of countries with only a reduced grid, the analysis determined cost savings from locally synergistic arrangements to serve not only the mines but also nearby towns and villages. For the group of countries with larger grids, the analysis laid out national and regional power-development opportunities for the mines. Finally, for the small number of countries with mature levels of power–mining integration, the analysis drew lessons and laid out options on how integration can be strengthened to create a win-win situation for all stakeholders.

Notes

1. Based on 2008 U.S. Energy Information Administration data.
2. P1 reserves are proven reserves (developed and undeveloped). P2 reserves are P1 (proven reserves) and probable reserves, thus proved and probable.
3. Two key exploration-related policies are first right to mine a discovery and ability to sell the discovery to a willing buyer.
4. Note that this figure was greater than 35 percent for all the countries except Liberia and Rwanda; by 2020 the figure for Liberia will likely be well over 50 percent.

The Power of the Mine • http://dx.doi.org/10.1596/978-1-4648-0292-8

5. Mineral and metal prices more or less stabilized in 2013 at well below their recent peaks, but were still more than double their 2003 prices for most commodities. It is difficult to predict the path of these prices as these global commodities are subject to many forces of supply and demand. The best predictor of mineral prices has been current price, and there is no evidence that forecasters can predict prices over the long period considered for mining investments. Daniel and others (2010) offers an interesting example of how inaccurate oil price forecasts can be.

6. See, for example, World Bank (2012b) on increasing local procurement in West Africa.

7. See, for example, analysis of resource corridors by Jourdan (2008).

8. Mining-rich Gabon is an outlier among SSA countries, exhibiting one of the region's highest electrification rates.

References

Andruleit, Harald, Hans Georg Babies, Jürgen Meßner, Sönke Rehder, Michael Schauer, and Sandro Schmidt. 2012. *Annual Report: Reserves, Resources and Availability of Energy Resources 2011.* Hannover, Germany: German Mineral Resources Agency.

Banerjee, Sudeshna G., Mikul Bhatia, Gabriela E. Azuela, Ivan Jaques, Ashok Sarkar, Elisa Portale, Irina Bushueva, Nicolina Angelou, and Javier G. Inon. 2013. *Global Tracking Framework.* Washington, DC: World Bank/Energy Sector Management Assistance Program and the International Energy Agency.

Bazilian, Morgan, Patrick Nussbaumer, Hans-Holger Rogner, Abeeku Brew-Hammond, Vivien Foster, Shonali Pachauri, Eric Williams, Mark Howells, Philippe Niyongabo, Lawrence Musaba, Brian Ó Gallachóir, Mark Radka, and Daniel M. Kammen. 2012. "Energy Access Scenarios to 2030 for the Power Sector in Sub-Saharan Africa." *Utilities Policy* 20 (2012): 1–16.

Collier, Paul. 2010. *The Plundered Planet: Why We Must—and How We Can—Manage Nature for Global Prosperity.* New York: Oxford University Press.

Daniel, Philip, Brenton Goldsworthy, Wojciech Maliszewski, Diego M. Puyo, and Alistair Watson. 2010. "Evaluating Fiscal Regimes for Resource Projects: An Example from Oil Development." In *The Taxation of Petroleum and Minerals: Principles, Problems and Practice*, edited by Philip Daniel, Michael Keen, and Charles McPherson, 187–240. London: Routledge.

ECA (Economic Consulting Associates). 2013. *Power and Mining in Africa.* Final Synthesis Report Submitted to the World Bank. London.

Foster, Vivien, and Cecilia Briceño-Garmendia. 2010. *Africa's Infrastructure: A Time for Transformation.* Africa Development Forum Series. Washington, DC: World Bank.

ICA (Infrastructure Consortium for Africa). 2011. *When the Power Comes: An Analysis of IPPS in Africa.* Tunis.

IFC (International Finance Corporation). 2011. "Geothermal Energy Scoping Study for East Africa." Prepared by GreenMax Capital Advisors, Washington, DC.

IJHD (*International Journal of Hydropower and Dams*). 2009. "World Atlas." Hydropower and Dams. Available through "World Hydro Potential and Development." Surrey, U.K.

IMF (International Monetary Fund). 2013. *World Economic Outlook, April 2013.* Washington, DC.

Jourdan, Paul. 2008. "A Resource-Based African Development Strategy." Prepared for African Development Bank, March 2008. http://www.partnership-africa.org/sites/default/files /RADS-A%20Resource-Based%20African%20Development%20Strategy.pdf.

Kumar, Arun, Tormod Schei, Alfred Ahenkorah, Rodolfo Caceres Rodriguez, Jean-Michel Devernay, Marcos Freitas, Douglas Hall, Ånund Killingtveit, and Zhiyu Liu. 2011. "Hydropower." In IPCC *Special Report on Renewable Energy Sources and Climate Change Mitigation*, edited by Ottmar Edenhofer, Ramon Pichs-Madruga, Youba Sokona, Kristin Seyboth, Patrick Matschoss, Susanne Kadner, Timm Zwickel, Patrick Eickemeier, Gerritt Hansen, Steffen Schlömer, and Christoph von Stechow, 437–96. Cambridge, U.K: Cambridge University Press.

SNL Metals Economics Group. 2013. "Worldwide Exploration Trends 2013." Halifax, Canada. http://www.metalseconomics.com/sites/default/files/uploads/PDFs/meg _wetbrochure2013.pdf.

UNDP (United Nations Development Programme). 2013. *Human Development Report 2013: The Rise of the South: Human Progress in a Diverse World*. New York. http://hdr .undp.org/en/reports/global/hdr2013/.

USGS (United States Geological Survey). 2013. "USGS Minerals Information." U.S. Department of Energy, Washington, DC. http://minerals.usgs.gov/minerals /pubs/commodity/.

Wilburn, D. R., and K. A. Stanley. 2013. "Exploration Review." United States Geological Survey, Reston, VA.

World Bank. 2012a. *Africa—Second Additional Financing for Southern African Power Market Project: Restructuring*. Washington, DC. http://documents.worldbank.org /curated/en/2012/06/16360805/africa-second-additional-financing-southern-african -power-market-project-restructuring.

———. 2012b. *Increasing Local Procurement by the Mining Industry in West Africa*. Report 66585-AFR, Washington, DC. http://documents.worldbank.org/curated /en/2012/01/15785549/increasing-local-procurement-mining-industry-west-africa -road-test-version.

———. 2013. *Enterprise Surveys*. Washington, DC. http://www.enterprisesurveys.org/.

CHAPTER 2

Mining Demand for Power

This chapter presents estimates of historical and projected power demand from Sub-Saharan Africa (SSA) mining activities over 2000–20, underpinned by the Africa Power–Mining Database 2014. Since 2000 the acceleration of SSA mining power demand has been driven by a range of countries beyond the historical mining powerhouse, South Africa. In some countries, mining power demand towers over nonmining demand, pointing to the need to integrate such demands in power planning.

Mines Require Enormous Amounts of Power

Mining operations require enormous amounts of power, depending on the type of mineral and even more on the amount of processing or beneficiation (box 2.1). The power required for fully beneficiating a mineral such as copper or aluminum smelting can be many times the amounts required for simple digging, crushing, and sorting. For copper, cobalt, and nickel, even the basic and intermediate processes are highly power intensive (figure 2.1; appendix B, table B.4).

Box 2.1 Mineral Beneficiation and Power Requirements

Mineral beneficiation refers to the processes that add value to the primary material produced by mining and extraction. Ore bodies nearly always contain a mixture of minerals, and it is necessary to separate desired minerals from undesired ones. For example, in most industrial gold mines, each ton of ore contains 2–5 grams of gold. Each successive level of processing permits the product to be sold at a higher price than the previous intermediate product or original raw material, thereby adding value. This study focuses only on the beneficiation processes that increase the concentration or purity of the mineral. It does not consider processes usually included in the manufacturing industry (such as car or jewelry making) as they rarely depend much, if at all, on the mine's location.

box continues next page

Box 2.1 Mineral Beneficiation and Power Requirements *(continued)*

Each mineral has its own particular processes, and many do not require undergoing all processes to be ready for use in manufacturing. The main processing stages are as follows:

• Extraction of the ore by digging, sorting, and crushing—commonly known as mining.
• Concentration of the mineral using various techniques, most of which are either gravity based or involve a chemical process (such as flotation) to separate out elements or use electric plates or magnets (electrostatic and magnetic separation) to separate waste from the mineral.
• Smelting of the mixture (concentrate) at high temperatures to burn off or separate undesired materials, in the form of slag, from the metal.
• Refining of the resulting output to further increase purity, often through electrolysis.

Figure 2.1 Cumulative Power Requirements for Mining Operations at Stages of Beneficiation

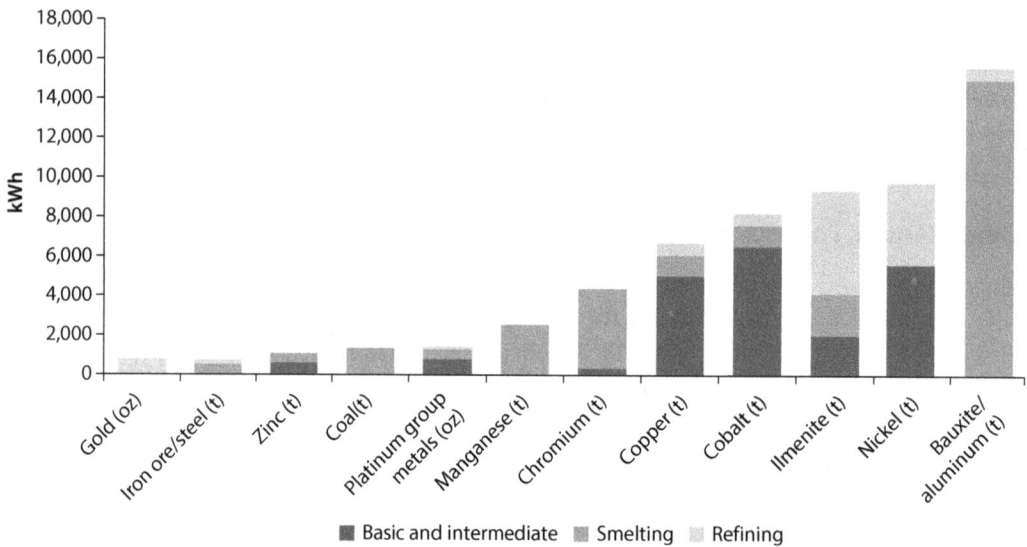

Source: USGS 2011.
Note: Refined bauxite is alumina.

Power costs are thus often an important part of a mining operation's cost structure, particularly if the mine must self-supply, meaning that generation is likely based on expensive diesel or heavy fuel oil (HFO). Power rarely constitutes less than 10 percent of mining operating costs and often exceeds 25 percent (table 2.1); these figures hold even though full beneficiation rarely occurs unless the mine has a relatively cheap or moderately priced source of grid power. To illustrate, if a typical grid supply cost of 10 cents per kilowatt-hour (kWh) or self-supply cost of 20 cents per kWh is considered, then power cost as a share of operating cost covers a wide range. Aluminum smelters in particular require very

Table 2.1 Installed Power Capacity Needed to Operate a Medium-Size Mine or Smelter for Selected Minerals

Mineral	Annual production (t)	Required power capacity, maximum beneficiation (MW)[a]	Electricity costs as a share of operating costs, maximum beneficiation (percent)[b]	
			10 cents per kWh	20 cents per kWh
Aluminum	200,000	443	117[c]	234
Bauxite	2,000,000	177	29	45
Coal	10,000,000	53	10	18
Cobalt	20,000	23	—	—
Copper	100,000	95	15	26
Diamond[d]	0.6	3	—	—
Gold (open-pit)	12	45	9	17
Gold (underground)	12	80	16	28
Ilmenite	300,000	15[e]	15	26
Iron ore/steel	3,000,000	338	16	28
Manganese	50,000[f]	121	11	20
Nickel	30,000	42	10	18
Platinum group metals	5.6	41	14	25
Uranium	1,814	46	30	46
Zinc	200,000	31	8	15

Source: World Bank.

Note: — = not available.

a. It is assumed that the plant has a capacity power of 80 percent.

b. Operating costs are calculated as 70 percent of the average July 2013 price of the metal. Operations extend to the refining or otherwise furthest stage in the process.

c. Due to the high power requirements per output value, aluminum smelters rarely operate unless power costs less than 3 cents per kWh.

d. A carat equals 0.2 grams; thus, there are 5 million carats in a ton of diamonds. The kWh requirement for diamonds includes separation.

e. This is for basic processing without refining. The amount of power varies greatly, depending on the end use.

f. Ferro-manganese.

cheap power: even at 10 cents per kWh, electricity costs will exceed the typical maximum operating costs, meaning that the smelter would not be viable. For manganese, nickel, and coal, electricity costs of 20 cents per kWh would not be prohibitive, as they constitute a relatively modest share of operating costs.

Except for Cameroon, Mozambique, South Africa, and Zambia, African countries process minerals only to the extent necessary for easy transport. South Africa's beneficiation dates back to the days of cheap coal and the self-sufficiency imposed by the apartheid era, while Zambia's beneficiation is driven by its cheap power, landlocked situation, and high transport costs. Nevertheless, mineral beneficiation is quite competitive, with smelters and refineries often located close to markets. The large rents are in the minerals themselves; if electricity is not priced at its true cost (that is, subsidized), beneficiation can even reduce value added.

The Power of the Mine • http://dx.doi.org/10.1596/978-1-4648-0292-8

A more important loss to a mining country caused by expensive power is that its small- and medium-size mining supply industries are at a competitive disadvantage with well-established countries (such as Australia, Canada, Chile, and South Africa). If the mining sector is to lead industrialization, establishing a strong mining supply industry is essential.

The grid capacity of many SSA countries, excluding South Africa, is limited: 22 out of 40 countries have fewer than 500 megawatts (MW), and 8 countries have fewer than 100 MW. For countries such as Liberia, Mauritania, and Sierra Leone, the power needed to operate just two medium-size mines (for example, for copper and gold) exceeds the capacity of the entire grid, so many countries' mines have to self-generate. Owing to the high energy costs of this option, they export relatively unprocessed minerals, as it would be unprofitable to beneficiate further. This situation is common across the continent, except in South Africa, which relies mainly on coal, and countries with large volumes of hydropower. Guinea is the prime example—none of its bauxite is brought to the final aluminum stage, and more than 90 percent is exported with only minimal processing. Most of the companies exploring and mining Guinea's enormous iron ore deposits plan to export the crushed ore without producing iron fines or pellets—and, of course, there are no known plans to produce steel.

Mining Demand for Power Could Triple to 23 GW by 2020

By 2020 mining power demand could triple from 2000 levels, reaching 23,443 MW. SSA's mining sector required 7,995 MW in 2000 and an incremental 7,130 MW in 2012 (figure 2.2a). Two scenarios project demand to 2020: high- and low-probability, depending on the stage in the project cycle and metal type, which, in turn, depend on prices, political stability, mining policies, and availability of infrastructure, including power (box 2.2). In 2020 power demand is expected to reach 155 percent of 2012 demand, when both high- and low-probability projects are included. This translates to a compound annual growth rate of 4.2 percent a year in the high-probability scenario and 5.4 percent when low-probability projects are included.

A handful of countries—primarily Cameroon, Guinea, Liberia, South Africa, the Democratic Republic of Congo, and Zambia—account for the difference between the high- and low-probability scenarios. All of these countries have many mining activities in the preliminary stages, which may or may not reach production status by 2020 (tables B.2 and B.3).

South Africa is by far SSA's most important mining country, constituting 70 percent of mining power demand in 2000 and 66 percent in 2012. But its contribution is forecast to decline to 56 percent if all high- and low-probability projects are realized. Even so, it will retain an overwhelming presence in SSA's mining landscape. So it is not surprising that Southern Africa stands out among regions for its mining power demand (figure 2.2b). In the high-probability scenario, this would mean a massive expansion in power demand of nearly 10,000 MW over 2000–20. Southern Africa remains the most important region, even when South

Figure 2.2 Mining Demand for Power

a. Over time

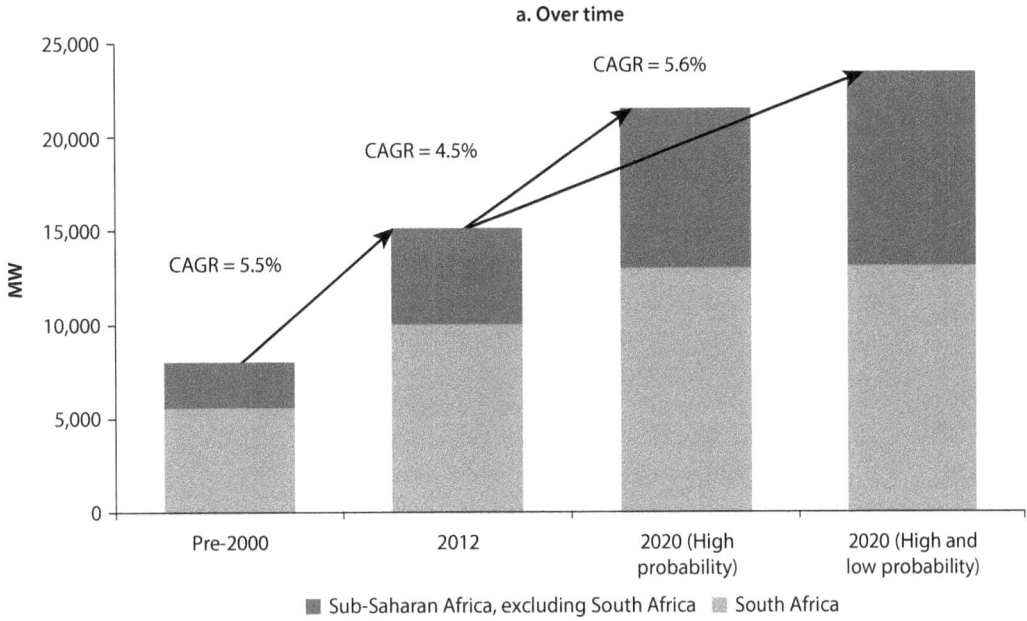

CAGR = 5.6%

CAGR = 4.5%

CAGR = 5.5%

■ Sub-Saharan Africa, excluding South Africa   ▨ South Africa

b. By region

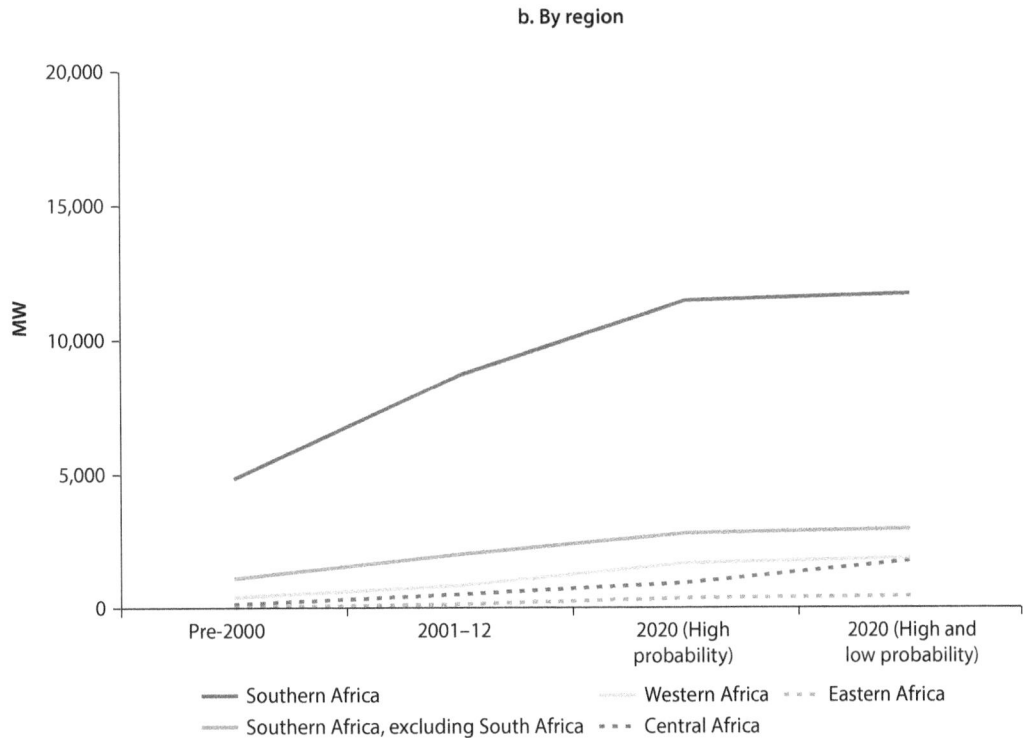

—— Southern Africa   ⋯⋯ Western Africa   - - - Eastern Africa

—— Southern Africa, excluding South Africa   ▪ ▪ ▪ Central Africa

Source: Africa Power–Mining Database 2014, World Bank, Washington, DC.
Note: CAGR = compound annual growth rate.

Box 2.2 Africa Power–Mining Database 2014—and Two Probabilities

The Africa Power–Mining Database 2014 draws on basic mining data from Infomine and the United States Geological Survey, annual reports, technical reports, feasibility studies, investor presentations, sustainability reports on property-owner websites or filed in public domains, and mining websites (Mining Weekly, Mining Journal, Mbendi, Mining-technology, and Miningmx).

This database, comprising 455 projects in 28 Sub-Saharan countries, with ore reserve value assessed at more than $250 million, collates publicly available and proprietary information and provides a panoramic view of projects operating in 2000–12, as well as anticipated demand in 2020. Projecting demand scenarios beyond 2020 is speculative, but 2020 demand can be considered a lower-bound estimate of demand in 2025 or 2030. The analysis is presented over three time frames: pre-2000, 2001–12, and 2020 (each containing the projects from the previous period, except for those that have closed).

The projected demand encompasses two scenarios, depending on the probability that the projects will be in production by 2020. The high-probability scenario includes all projects in prefeasibility, feasibility, development, and producer stages for nonprecious metals and includes projects in prospect, exploration, and advanced exploration stages as well for precious metals (for example, gold, rare earth metals, and diamonds). The low-probability scenario includes most projects in prospect, exploration, and advanced exploration stages for nonprecious metals. Projects in temporary suspension are placed in the low-probability scenario.

Note: Details on the database are available in table B.1.

Africa is excluded, followed by Central Africa and West Africa. East African mines have by far the smallest demand for power.

While South African mining will add sizable demand for power, with projected annual growth of 3.5 percent in 2012–20, the expected growth for the region's other countries is more impressive. In fact, the compound annual growth rate (CAGR) could reach 9.2 percent if all projects in the two scenarios are realized in the other SSA countries. Since 2000 the top 10 countries with the highest mining power demand have remained largely unchanged, except for their relative rankings (figure 2.3). Mozambique and Zambia retain the top two spots over the period. If its enormous potential for low-probability projects is realized, Cameroon will become one of the top countries for mining power demand.

Power Demand from Mines Is Concentrated in a Group of Metals

For 2000–20, aluminum leads the group, followed by copper, platinum group metals (PGM), chromium, and gold (figure 2.4a). For aluminum, however, there has been little new activity since 2000, and power demand will only grow in 2020 if low-probability projects are realized. The largest absolute increases

Figure 2.3 Mining Power Demand for Top 10 Countries (Excluding South Africa)

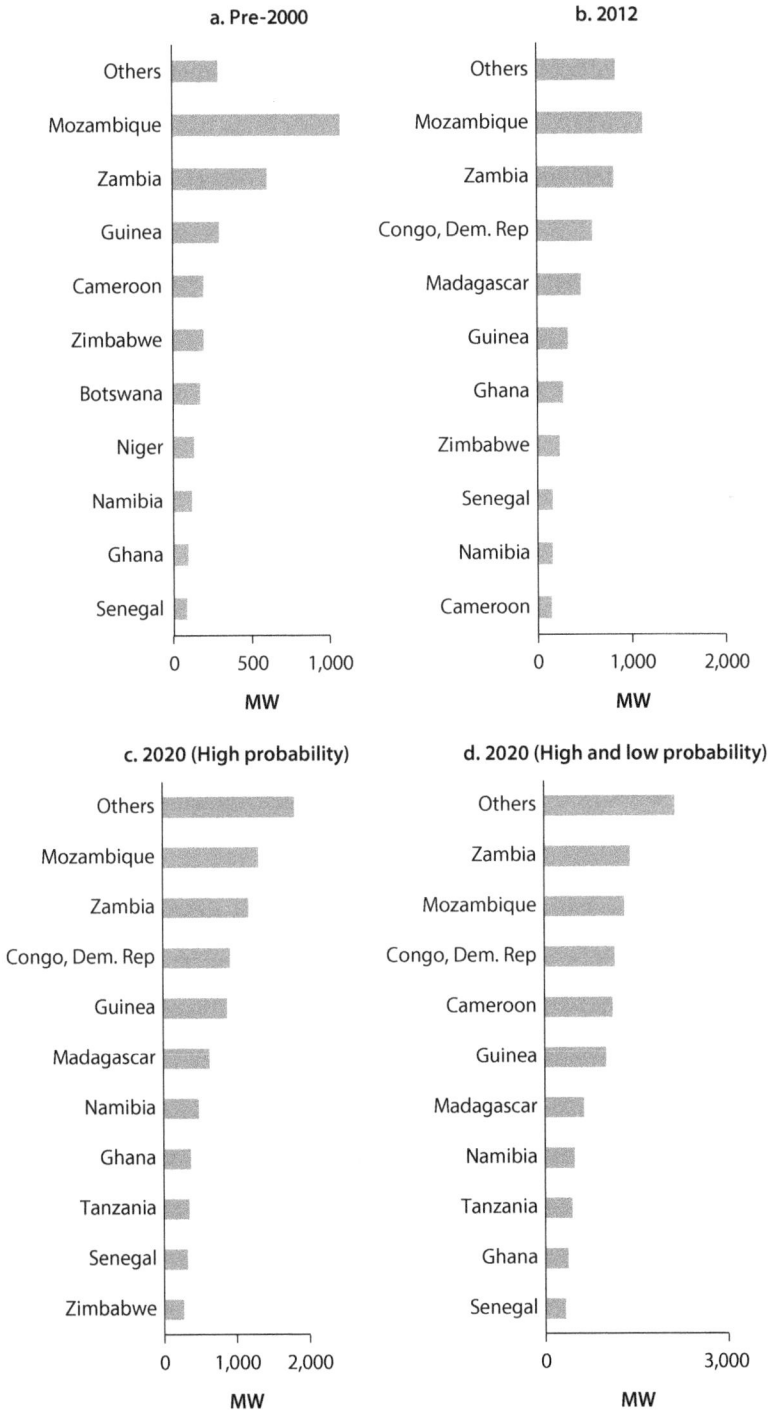

a. Pre-2000

b. 2012

c. 2020 (High probability)

d. 2020 (High and low probability)

Source: Africa Power–Mining Database 2014, World Bank, Washington, DC.

over the period will be for copper and PGM, which will increase demand by 2,150 MW and 2,010 MW, respectively, when high- and low-probability projects are included. Close behind is iron ore, which is expected to require an extra 1,486 MW of power between 2000 and 2020. The most substantial growth rate increases are expected for iron ore, nickel, PGM, silver, and zirconium. These last three commodities are found mainly in South Africa. Excluding projects in South Africa, SSA will experience larger increases in iron ore, at 31 percent, and gold, at 8 percent. But the primary driver of future production of these minerals will be based on the most important, yet most unpredictable, variable—the minerals' future market price. That the largest increases are associated with minerals used in large-scale infrastructure development is no coincidence; emerging economies continue to demand more of these commodities as they develop.

The smelting processes are the most power intensive and have contributed the most to mining power demand. For all periods, refining and smelting together account for more than three-quarters of power demand (figure 2.4b).

Figure 2.4 Mining Demand for Power

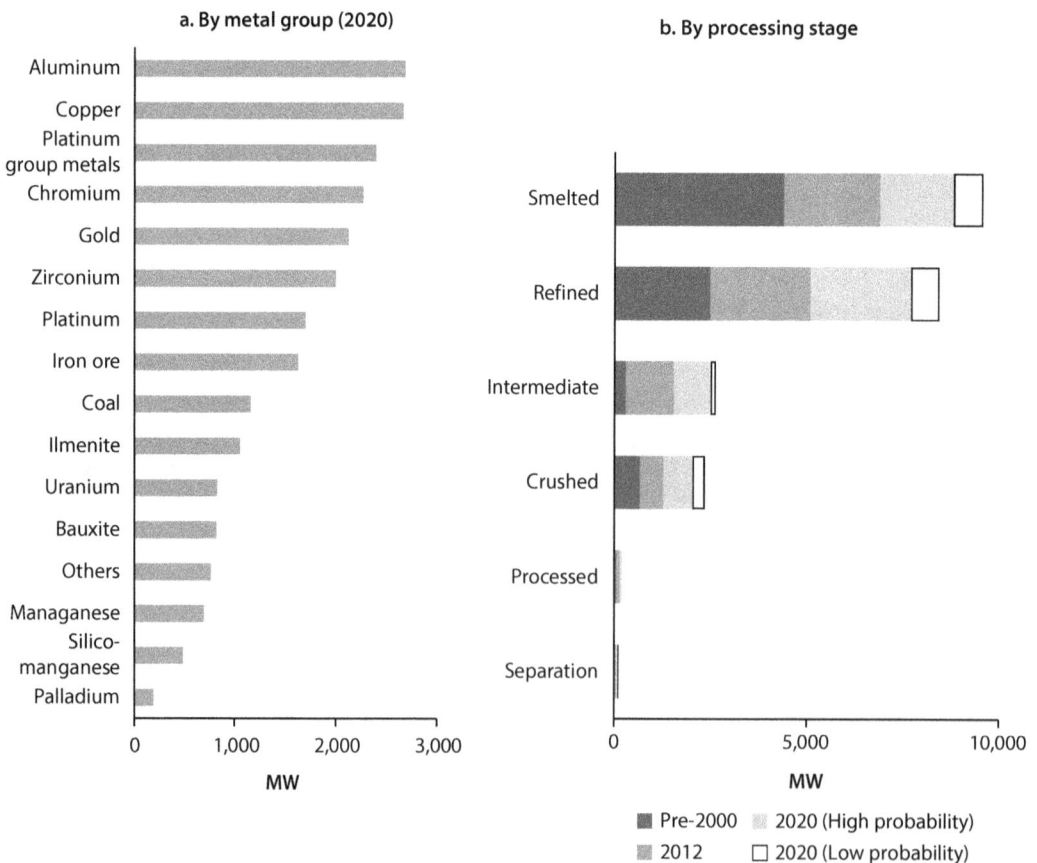

Source: Africa Power–Mining Database 2014, World Bank, Washington, DC.

Starting from a low base, the growth profile of separation and crushing activities—relatively less power-intensive steps—are envisaged to increase annually by 9 percent and 8 percent, respectively, over 2012–20.[1]

The annual energy consumption emanating from this demand can be from a small group of projects, particularly for such minerals as silico-manganese and aluminum. On the other hand, copper, which reports the highest annual consumption after aluminum, is associated with a large number of projects; so the average annual consumption is relatively lower. Not surprisingly, therefore, smelting reports the largest average annual consumption. By comparison, the stages of crushed, intermediate, and separation are much less power intensive (see figure 2.4b).

Mining Demand for Power Will Overwhelm Nonmining Demand for a Handful of Countries

For a handful of countries, mining power demand is quite large relative to non-mining demand, and will remain so for the rest of the decade. At the continental level, mining demand constitutes about 24 percent of 2012 nonmining demand, and is expected to rise to 30 percent of that demand by 2020, when high- and low-probability projects are considered (figure 2.5a). Tanzania aside, no country explicitly accounts for mining demand in their power sector master plans. Mining demand will be overwhelming in Guinea, Liberia, and Mozambique (figure 2.5b), where it is set to exceed nonmining demand by 2020.

Figure 2.5 Comparison of Mining and Nonmining Demand for Power

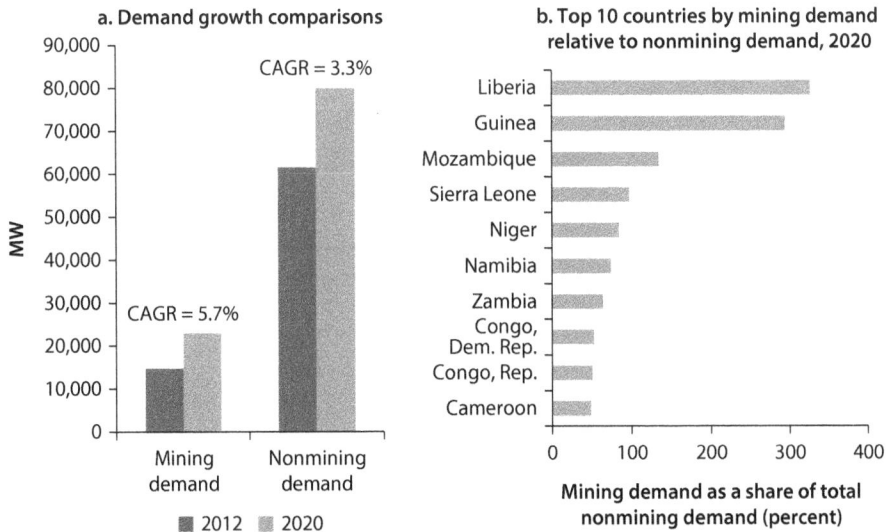

Sources: Africa Power–Mining Database 2014, World Bank, Washington, DC; Economic Consulting Associates analysis.
Note: Panel a includes 27 countries for which both mining demand and total demand figures are available.
CAGR = compound annual growth rate.

The Power of the Mine • http://dx.doi.org/10.1596/978-1-4648-0292-8

By 2020 mining power demand, now equivalent to about one-quarter of available grid supply, could climb to 35 percent of available 2012 supply, assuming both high- and low-growth scenarios. Averages mask large differences across countries: for six countries in 2012, mining power demand was equivalent to more than half of grid supply and more than three-quarters in three countries. Mining power demand relative to grid supply is projected to rise dramatically in Cameroon, Guinea, and Liberia, suggesting an urgent need to expand generation capacity and for governments to work toward incorporating demand into power planning. It is not hard to envisage mining power demand crowding out demand from less well-endowed households and small industrial customers. In countries such as Zimbabwe, mining companies already pay a premium to guarantee "first access" to power.

Note

1. In 2013 the Democratic Republic of Congo, Ethiopia, South Africa, and Zimbabwe took steps to make domestic beneficiation of certain mineral products mandatory; if these policies are effective, power demand from the mining sector could rise considerably beyond what is reported in this study.

Reference

USGS (United States Geological Survey). 2011. "Estimates of Electricity Requirements for the Recovery of Mineral Commodities, with Examples Applied to Sub-Saharan Africa." U.S. Department of Energy, Washington, DC.

CHAPTER 3

Mine Power-Sourcing Arrangements

This chapter develops a typology of Sub-Saharan Africa (SSA) mine power-sourcing arrangements, focusing on trends and causal factors. Traditionally, grid supply has been the most common option; however, the future is frontier terrain, with intermediate options evolving to meet demand. Self-supply, chosen by a small number of projects in 2000, will also experience a dramatic upward trend to 2020. The choice of options correlates with country grid tariffs, supply reliability, and fuel mix.

A Typology of Arrangements

The spectrum of power-sourcing arrangements in SSA ranges from grid supply to self-supply. Traditionally, mines have sourced power from the national grid because of its secure and reliable supply and cost effectiveness against self-generation. The drawback is that mines have sometimes continued relying on the grid model after its supply attributes have deteriorated. In the self-supply model, mines produce their own power because of the high costs of extending transmission and distribution networks to the mining site, poor supply security from the grid, or high tariffs where the grid is powered by diesel or heavy fuel oil (HFO). Some mines are willing to pay a higher cost for self-supply to ensure reliability rather than source from the grid.

Six intermediate arrangements reflect the novel ways in which mines, utilities, and governments are coming together for mutual benefit. The mines need reliable power, the utilities need revenue-generating anchor consumers and mining load for national power system development, and governments need to improve the utilities' commercial viability for broader power sector development.

These six intermediate arrangements lead to synergies from operational cost effectiveness, greater availability of electricity, accelerated expansion, and more robust transmission across SSA. The arrangements are as follows: self-supply and corporate social responsibility (CSR), self-supply and sell to the grid, grid supply and self-supply backup, mines sell collectively to the grid, mines invest in the grid, and mines serve as anchor demand for an independent power producer (IPP) (table 3.1). There is a continuum of interaction with the grid from the second option on. While the final arrangement is one of the most relevant for power–mining integration, South Africa is so far the only country in the region to have used it. Ghana is the only other SSA country demonstrating collective behavior—mines have joined to create a coordinated investment and sell the excess to the grid.

Beyond these clearly defined intermediate options, combinations of arrangements—or transitional arrangements—can evolve. For example, a mine may choose to use self-supply as its primary power source and also look for possible developers to source from an IPP. Such a project would include both arrangements over the mine's lifetime. But in practice the mine would gradually transition from one arrangement to the other; by 2020 it is expected that the mine will have shifted to the latter option.

The Share of Grid Supply in Projects Has Declined

Grid supply dominates projects' power-sourcing arrangements, although its share has declined from 60 percent to 48 percent over 2000–20, suggesting that mining companies are either using self-supply or choosing intermediate arrangements to meet their power needs. In recent years these intermediate options have registered an impressive rise, from 43 projects before 2000 to 123 projects in the 2020 high-probability scenario and 132 projects when low-probability projects are included. Their contribution to total projects has remained stable. The self-supply option has seen the most dramatic rise, from only 6 percent of projects before 2000 to 18 percent in 2020. In compound annual growth rate (CAGR), self-supply projects rose the fastest, at 11.5 percent, followed by intermediate options, at 5.8 percent, and grid supply, at 4.7 percent (figure 3.1). Despite the steep rise for the self-supply option, the model remains a minority among power-sourcing arrangements in 2020 (figure 3.2).

Among the intermediate options, three trends emerge. First, the two most commonly chosen options over the 20-year period have been "self-supply and sell to the grid" and "mines invest in the grid." Both options require the mines and utilities to interact in ways that involve investment. In the first, the mines provide for themselves and sell the excess to the grid, which may require investments in transmission and distribution networks, with a concomitant impact on neighboring areas and overall power systems. Similarly, in the second, the mines and utilities collaborate to invest in existing and new assets. Second, the prevalence of transitional arrangements stays roughly similar, suggesting that power sourcing is dynamic. The mines respond nimbly to changing situations. Before 2000, transitional projects constituted 23 percent of projects choosing intermediate options, rising slightly to 24 percent in 2020. Third, in 2020 not only will more projects rely primarily on self-supply, but many projects will also use self-supply as backup, even when connected to the grid. This option is set to experience a sharp rise.

Table 3.1 Typology of Power-Sourcing Arrangements

	Self-supply	Self-supply + CSR	Intermediate options					Grid supply
			Self-supply + sell to the grid	Grid supply + self-supply backup	Mines sell collectively to the grid	Mines invest in the grid	Mines serve as anchor demand for IPP	
Description	Mine produces its own power for its own needs	Mine provides power to community through mini-grids or off-grid solutions	Mine produces its own power and sells excess power to the grid	The mine is first connected to the grid and is moving into own-generation when more economical	Coordinated investment by a group of mines, producers, and users in one large power plant off-site connected to the grid	Mine invests with government in new, or in the upgrading of, power assets under different arrangements	Mine buys power from an IPP and serves as an anchor customer	Mine does not produce any power, but buys 100% from the grid
Main generation drivers	Diesel, HFO	Diesel, HFO	Coal, Gas, Hydro	Diesel, HFO	Diesel, HFO, Solar	Hydro, Gas	Any	Any
Presence	Mali and Guinea (hydro) Sierra Leone and Liberia (oil)	Guinea, Madagascar	Zimbabwe Mozambique, Cameroon	Congo, Dem. Rep. Tanzania	Ghana	Niger Congo, Dem. Rep.	South Africa	Mozambique Zambia

Source: Africa Power–Mining Database 2014, World Bank, Washington, DC.
Note: CSR = corporate social responsibility; HFO = heavy fuel oil; IPP = independent power producer.

Figure 3.1 Evolution of Power-Sourcing Arrangements

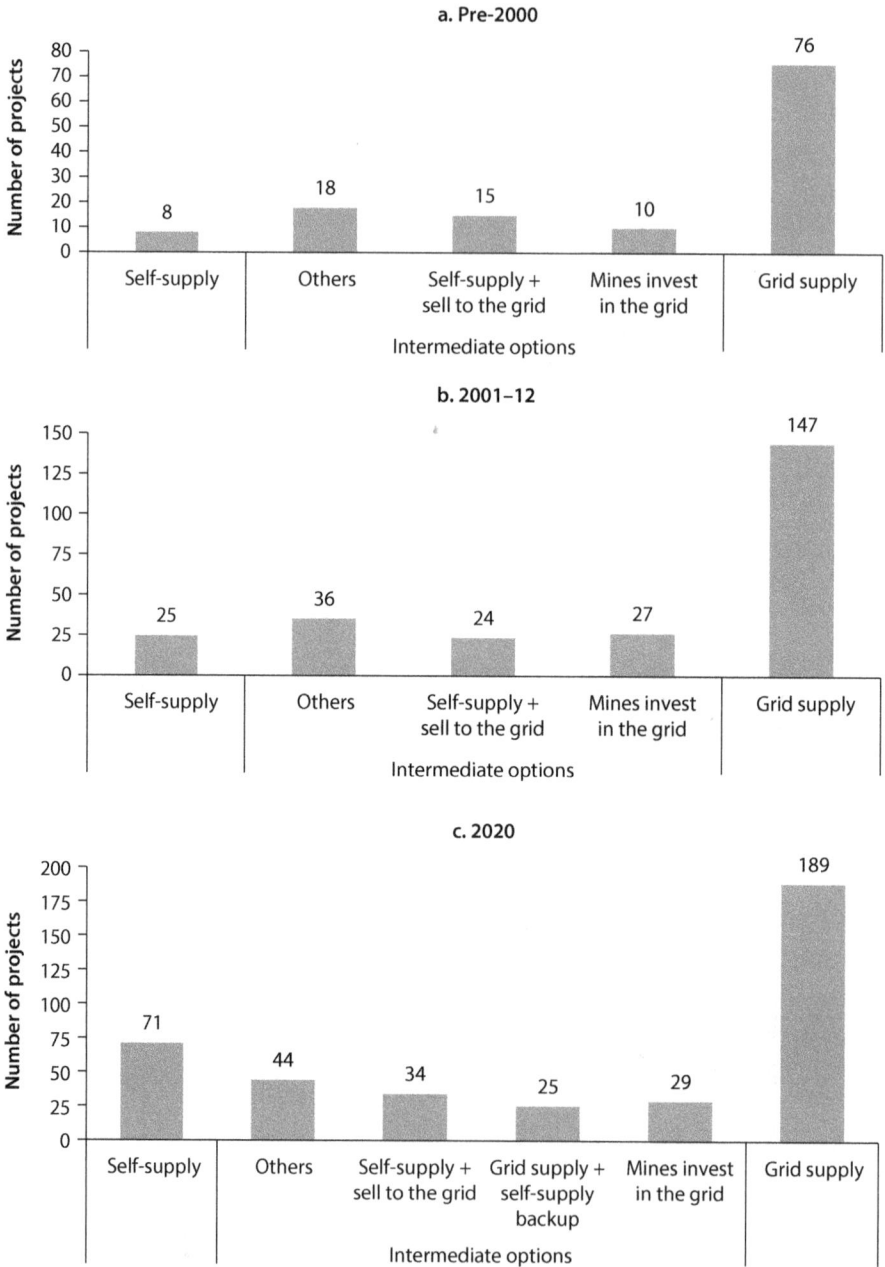

a. Pre-2000

b. 2001–12

c. 2020

figure continues next page

Figure 3.1 Evolution of Power-Sourcing Arrangements *(continued)*

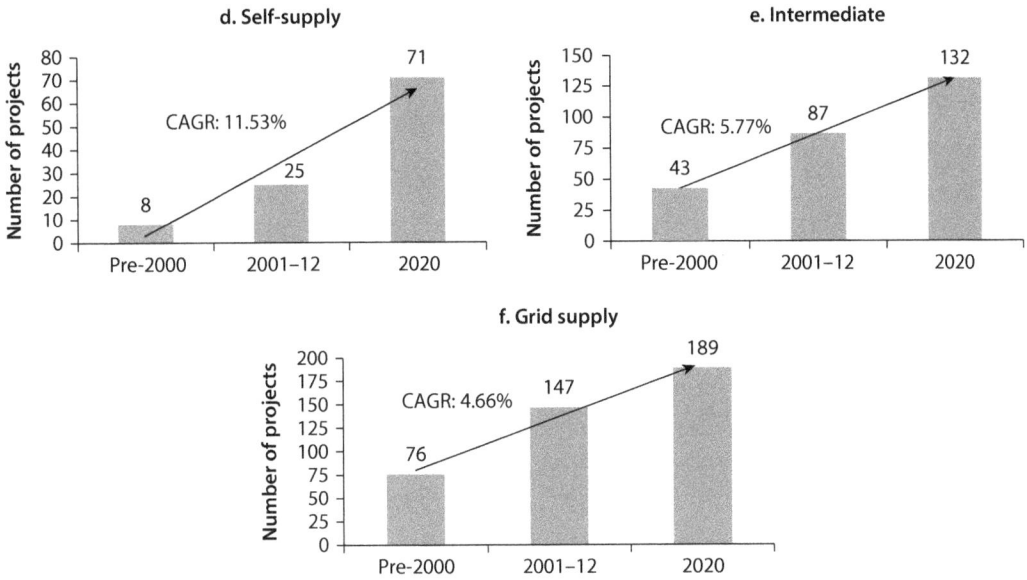

d. Self-supply

e. Intermediate

f. Grid supply

Source: Africa Power–Mining Database 2014, World Bank, Washington, DC.
Note: Includes both high- and low-probability projects. CAGR = compound annual growth rate.

Figure 3.2 Power-Sourcing Options

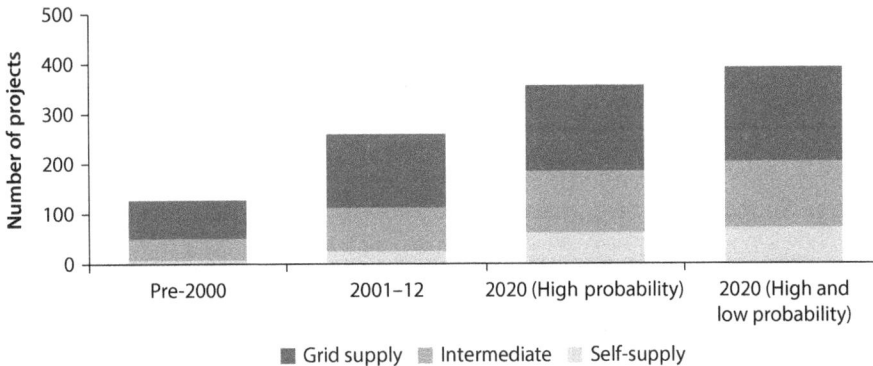

■ Grid supply   ■ Intermediate   ▨ Self-supply

Source: Africa Power–Mining Database 2014, World Bank, Washington, DC.

Intermediate Options Report the Largest Rise in Annual Consumption

Due to the rising number of mining projects across all arrangements, annual energy consumption grows for all options (figure 3.3a). Between 2000 and 2020, intermediate options show the highest growth, with a CAGR of 7.8 percent, followed by self-supply and grid supply, at 5 percent each. But the average annual energy consumption for self-supply projects declines by 5.8 percent. Intermediate options report a rise of 1.9 percent, while grid supply remains neutral (figure 3.3b).

Figure 3.3 Energy Consumption

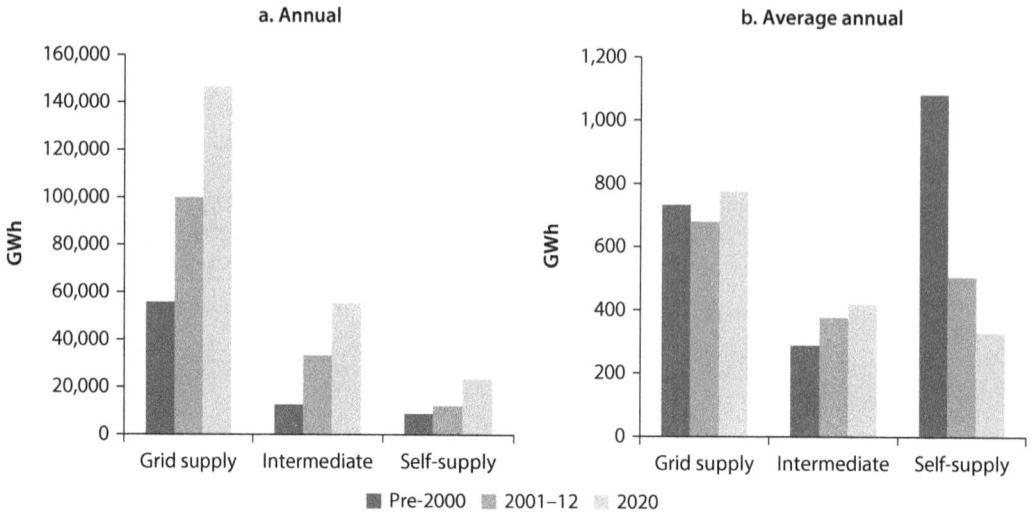

Source: Africa Power–Mining Database 2014, World Bank, Washington, DC.

This means that self-supply supports a greater number of smaller projects. Larger projects are supported by the grid or intermediate options (tables B.5 and B.6).

For both high- and low-probability projects, South African mines will drive grid supply in 2020, accounting for at least 70 percent of grid-based power consumption, followed by Zambia and Mozambique (figure 3.4a). For the intermediate options, Cameroon and the Democratic Republic of Congo lead energy consumption after South Africa (figure 3.4b). For self-supply, Guinea, Liberia, Mali, and Tanzania loom large on the landscape after South Africa (figure 3.4c). After South Africa, Guinea will be the main driver behind this arrangement in 2020.

Power-Sourcing Arrangements Have Evolved in SSA Regions

The evolution of SSA's mine power-sourcing landscape reveals regional differences. Before 2000 Central Africa's power-sourcing arrangements depended heavily on grid supply, although intermediate options subsequently became more relevant. The main intermediate arrangement is "mines invest in the grid." In the Democratic Republic of Congo, mines collaborate with the public utility, Société Nationale d'Electricité (SNEL) (figure 3.5a).

In East Africa, intermediate options dominated before 2000, while grid supply and self-supply options will have emerged by 2020. Self-supply is driven by Eritrea's Bisha copper project. Because that country's grid is based mainly on fuel oil, the higher tariff is nudging mines to turn to self-generation. Regional grid supply is driven by Kenya, notably the Kwale ilmenite mine. Hydro dominates Kenya's generation, so mining companies prefer grid-sourced power. The majority of East Africa's mines that use intermediate arrangements are located in Tanzania. Because of that country's limited transmission network, mines

Figure 3.4 Country Composition of Annual Energy Consumption, 2020

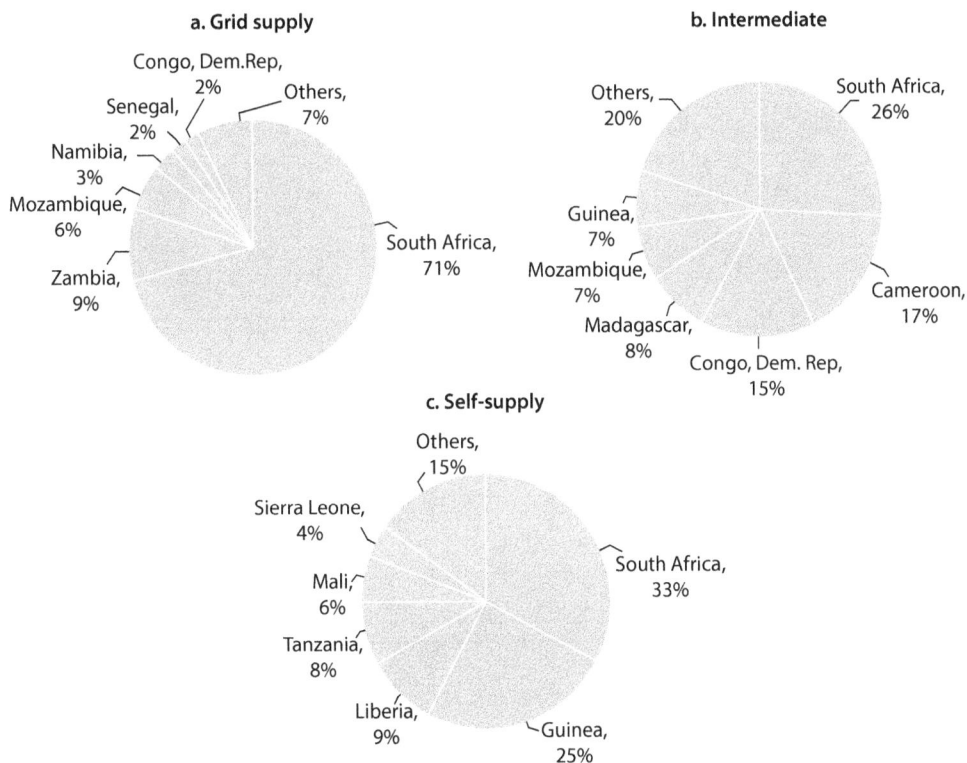

Source: Africa Power–Mining Database 2014, World Bank, Washington, DC.

either invest with the utility to upgrade the network or increase their reliance on self-generation (figure 3.5b).

Of all SSA regions, West Africa has the highest incidence of self-supply, particularly since the largest power-intensive projects are in countries where energy resources are unexploited and the grid is either below capacity or nascent (for example, Guinea, Liberia, and Sierra Leone). So it is more cost effective for mines to rely on self-generation. Prior to 2012 Ghana's Volta Aluminum Company Limited (VALCO) smelter was the largest driver of grid supply; in 2013 Senegal's Grande Côte ilmenite mine became the largest. Over 2000–20 Niger is the largest user of the "mines invest in the grid" option, but Sierra Leone's Tonkolili project, for which African Minerals is expanding the Bumbuna dam, is expected to become a large consumer under this arrangement by 2020. Guinea is the largest driver of the self-supply and CSR option for all time periods; that country has legacy systems and agreements in place that allow its mines to produce electric power for local communities (figure 3.5c).

Compared with other SSA regions, Southern Africa is fairly well integrated. The surge in the option of "self-supply and sell to the grid" is due to the region's increased development of coal projects, notably in Mozambique and South Africa. These projects, almost by definition, sell extra capacity to the grid.

Figure 3.5 Regional Power-Sourcing Arrangements over Time
Percent

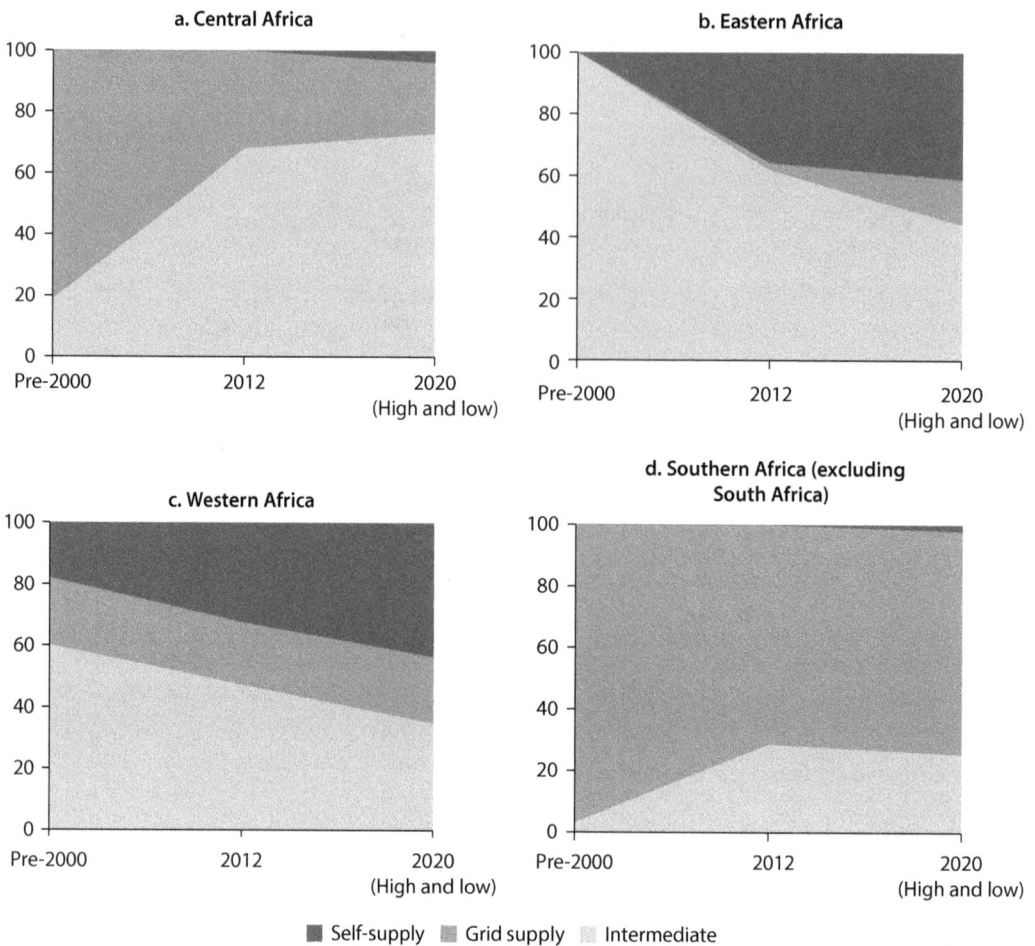

a. Central Africa

b. Eastern Africa

c. Western Africa

d. Southern Africa (excluding South Africa)

■ Self-supply ▨ Grid supply ▨ Intermediate

Source: Africa Power–Mining Database 2014, World Bank, Washington, DC.

Over 2000–20 the option of "mines invest in the grid" has been used mainly in Botswana, where mines frequently work with the public utility, Botswana Power Corporation, and in Mozambique, where Kenmare's Moma ilmenite mine works with the utility, Electricidade de Moçambique (EDM), to ensure the mine's access to the grid (figure 3.5d).

Self-Supply Imposes a Heavy Burden

Among all options, self-supply is the most expensive; it averages 23 cents per kilowatt-hour (kWh) and can reach 29 cents per kWh (figure 3.6). All self-supply facilities are fueled primarily by diesel, followed by HFO; these are also large sources of carbon emissions. The grid option is the cheapest, averaging about

Figure 3.6 Range of Energy Costs

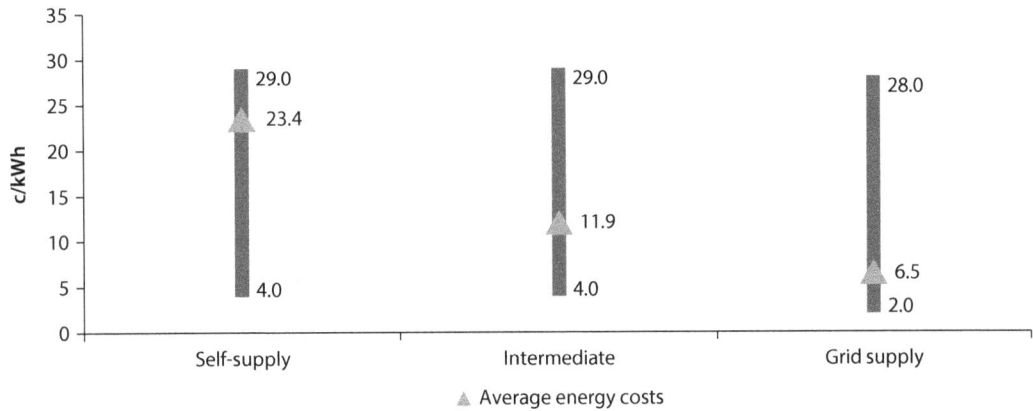

Source: Africa Power–Mining Database 2014, World Bank, Washington, DC.

6 cents per kWh. The minimum grid tariff of 2 cents per kWh is found in Lesotho's hydropower-fueled diamond projects. For intermediate options, the average falls somewhere in between, at 12 cents per kWh, and can run as low as 5 cents per kWh (for example, the transitional arrangement of Angola's Catoca diamond mine from "self-supply and CSR" to "mines serve as anchor demand for IPP"), or average 6 cents per kWh for the options of "mines invest in the grid" and "mines serve as anchor demand for IPP." In other transitional arrangements, forming part of self-supply generation, the average price can reach 28 cents per kWh (figure 3.6).

SSA energy costs since 2000 also point to the size of resources spent on each of these arrangements. The utilities have recouped about $75 billion from mines, followed by intermediate options, at $63 billion. But spending on self-supply arrangements has represented a loss to both the utilities and the country—the situation will worsen as mines choose more often to self-supply to meet their power needs. Self-supply spending, at $14 billion in 2000–12, is envisaged to climb to $22 billion in 2013–20. Bringing more of these resources within the fold of grid supply or intermediate options can benefit a larger group of stakeholders beyond the mines.

For power-sourcing arrangements involving grid-based partnerships (mines sell collectively to the grid, mines invest in the grid, mines serve as anchor demand for IPP and grid supply), by 2020 coal-based generation will account for 69 percent of supply in the high-probability scenario and 67 percent when low-probability projects are included. Hydropower generation will account for about 27–29 percent, depending on high- and low-probability, with the remaining 4 percent fueled by oil (figure 3.7a). The large share of coal-powered grid generation will be driven by South Africa, at about 11,064 megawatts (MW) and 11,887 MW in the high- and low-probability scenarios, respectively. Excluding South Africa's installed capacity, clean energy resources will play an important role; mining power requirements could help to unlock the clean energy potential, at 4,490 MW and 5,000 MW in the high- and low-probability scenarios, respectively (figure 3.7b).

Figure 3.7 Comparison of Main Power-Generation Sources in Collective Arrangements

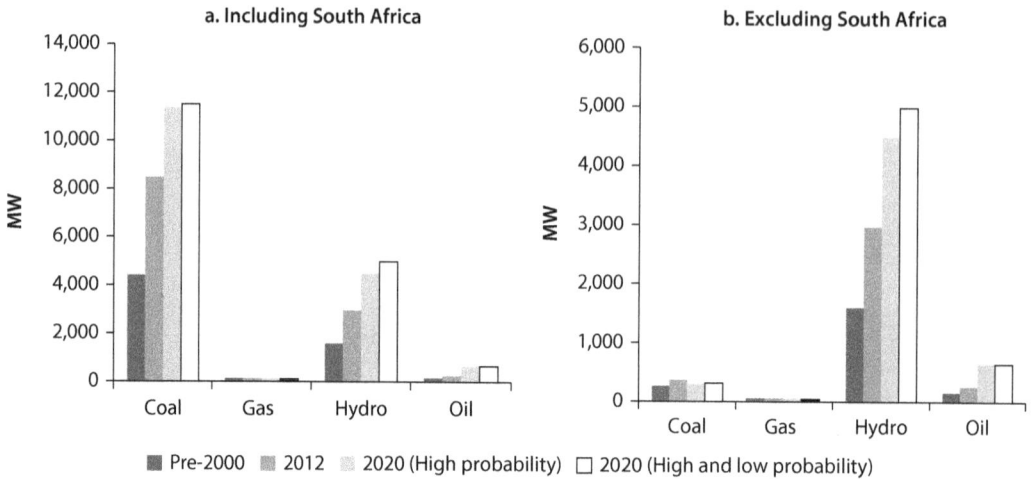

a. Including South Africa

b. Excluding South Africa

Pre-2000   ▨ 2012   ▨ 2020 (High probability)   ☐ 2020 (High and low probability)

Source: Africa Power–Mining Database 2014, World Bank, Washington, DC.

Mines have invested substantial resources in building power generation capacity, and these investments are slated to grow. Over 2000–12 about $1.3 billion was invested in about 1,590 MW of generation capacity in arrangements that involved some form of self-supply (self-supply; self-supply + CSR; self-supply + sell to the grid; grid supply + self-supply backup). Looking forward to 2020, around 10,260 MW are expected to be added to meet mining demand. Of this, arrangements that involved some form of self-supply will be 1,753 MW in the high-probability scenario and 3,061 MW when low-probability projects are included. This will amount to $1.4 billion to $3.3 billion, 21–30 percent from the total generation capacity required to meet 2020 mining demand. Most of this capacity will be thermal (diesel and HFO) generation capacity. The remainder of the required generation capacity of about 6,650 MW in the high-probability scenario and 7,195 MW when low-probability projects are included is expected to be met by collective arrangements (mines sell collectively to grid; mines invest in grid; mines serve as anchor demand for IPP) and grid supply.

Three Main Factors Determine Power-Sourcing Arrangements

First, a country's primary energy source is important for adequate and reliable supply. Hydropower-based grids will incentivize integration; that is, mines are either connected to the grid or seek an arrangement with the utilities (figure 3.8a). When energy from the grid is inexpensive, mines will seek this integration with a certain level of their own backup power generation. Gas-based grids tend to have the effect of hydropower-based grids, but are still quite few. The high incidence of the option whereby mines self-supply and sell excess to the grid in hydropower-dominant countries is due to two massive coal

Figure 3.8 Power Arrangements of Mining Firms Relative to the Three Main Factors

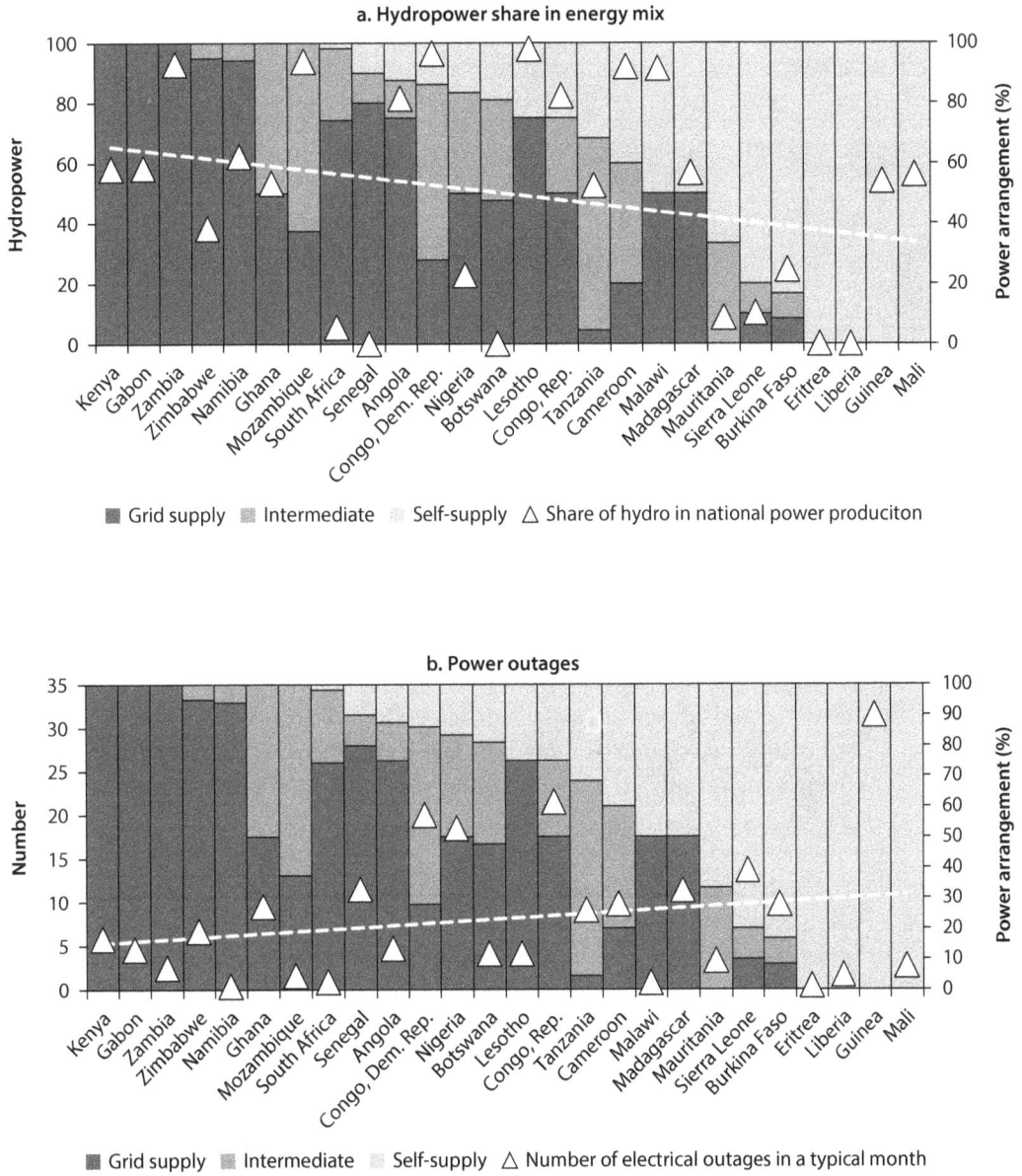

a. Hydropower share in energy mix

■ Grid supply ▦ Intermediate ⬚ Self-supply △ Share of hydro in national power produciton

b. Power outages

■ Grid supply ▦ Intermediate ⬚ Self-supply △ Number of electrical outages in a typical month

figure continues next page

Figure 3.8 Power Arrangements of Mining Firms Relative to the Three Main Factors *(continued)*

c. Electricity tariffs

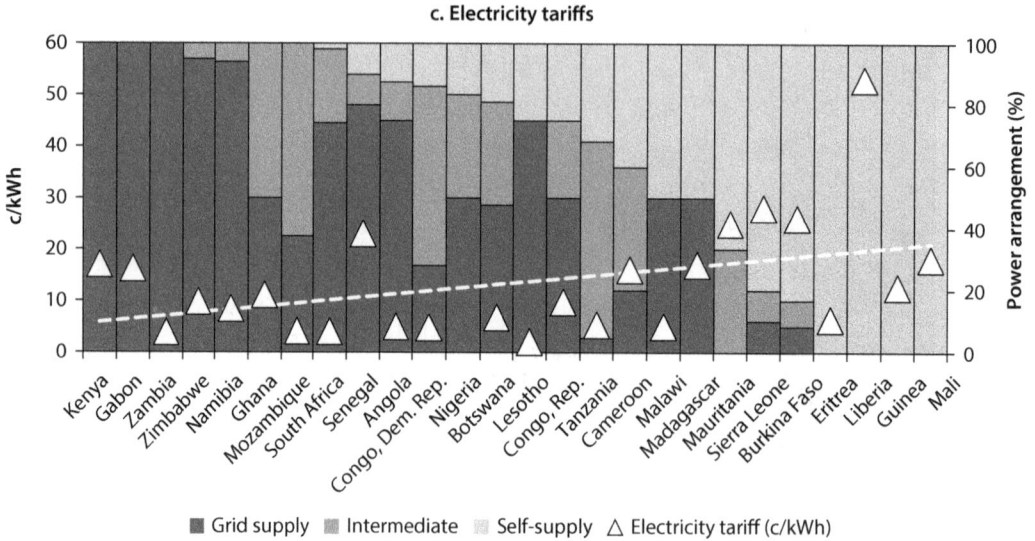

■ Grid supply   ▩ Intermediate   ▨ Self-supply   △ Electricity tariff (c/kWh)

Source: Africa Power–Mining Database 2014, World Bank, Washington, DC; Eberhard et al. 2011.

mines in Mozambique, Vale's Moatize and Rio Tinto's Benga, which are planning to sell electricity to the grid from their coal-fired power plants.

Second, the supply–demand imbalances and resulting power shortages also push certain arrangements. Low grid supply reliability is associated with mines moving toward self-supply. For example, grid-connected projects in South Africa are increasingly moving into self-supply when this is economical. In the past, all of the country's mines were connected to the grid, but the energy crisis drove them to transition to self-supply. Similarly, in Tanzania, gold mines that previously chose to connect to the grid are transitioning to self-supply, given the excessive outages caused by transmission-line bottlenecks (figure 3.8b).

Third, cost differentials between grid tariffs and self-supply costs play a significant role (figure 3.8c). Self-supply is present in two situations: HFO-based grids, particularly in Liberia and Sierra Leone, where mines self-supply at a lower cost than the grid tariff, and hydropower-based countries, especially Guinea, that suffer from high outages and where mines self-supply at a higher cost than the utilities. Given the hidden cost of outages, self-supply remains cost effective. The situation where mines invest with government in existing or new power assets is seen mainly in countries with more frequent power outages and hydro-based grids. These conditions largely justify mines co-investing with utilities in power infrastructure: mines expect to build a stronger grid to deliver stable and cheap electricity in the medium to long run.

Mines are more likely to self-supply when electricity tariffs are higher. The price of electricity depends heavily on the relative abundance of low-cost energy available for power generation. Hydropower is the main source of grid power for a sample of SSA countries, followed by oil, coal, and gas. Hydropower has significant economies of scale and can generally deliver the lowest cost electricity on a levelized cost basis, while oil-based supply is the most expensive. So the average price of electricity and percentage of self-supply are positively correlated with the share of oil and negatively correlated with the share of hydropower in electricity generation, although tariffs are by no means the only concern of mining companies (see chapter 5). As the number of monthly outages rises, a mine is more likely to choose options other than grid supply to meet its electricity needs. Beyond these factors, the power arrangement depends on access to the grid. This may explain why mines in Mali, characterized by fairly low tariffs and infrequent outages, use the self-supply model; that is, for mines in remote areas, grid connection may be uneconomical, regardless of the fuel.

References

Eberhard, Anton, Orvika Rosnes, Maria Shkaratan, and Haakon Vennemo. 2011. *Africa's Power Infrastructure: Investment, Integration, Efficiency.* Directions in Development. Washington, DC: World Bank.

World Bank. 2013. *Enterprise Surveys.* Washington, DC. http://www.enterprisesurveys.org/.

CHAPTER 4

Opportunities and Lessons for Power–Mining Integration

This chapter reviews the potential of power–mining synergies in eight countries representative of the various integration stages found in Sub-Saharan Africa (SSA). A first analysis simulates power–mining integration opportunities in four countries. This exercise aims to illustrate a range of innovative options to 2020, given the limitations of longer term predictions. In all of the countries examined, there are possibilities for mines to contribute to national and, in some cases, regional power sector development, which can be explored in countries with similar contexts. Second, lessons are presented from countries where mining and power have achieved enough integration to provide a reference in identifying the benefits and risks.

This chapter is complemented by three sets of appendixes—maps (appendix A), data (appendix B, tables B.7–B.10), and methodology and assumptions (appendix C).

Power–Mining Overview

The selection of countries for this analysis illustrates the ways power–mining integration can occur in SSA, taking into consideration utility performance, opportunities for regional integration, mining's contribution to the country's economy, and the potential of mining power demand to unlock clean and large domestic energy resources. The selected countries are Cameroon, the Democratic Republic of Congo, Ghana, Guinea, Mauritania, Mozambique, Tanzania, and Zambia.

Mining contributes substantially to these countries' economies except for Cameroon, which relies on mining as a source of foreign currency. Mining also contributes a major share of gross domestic product (GDP), the highest of which is in Mauritania, at 37 percent, followed by Guinea, at 18 percent, and the Democratic Republic of Congo, at 12 percent. In these three countries, along with Mozambique, mining's share of GDP is most likely to rise, given the extent of current and planned mining activities.

These eight countries are also well endowed with considerable energy resources, particularly gas and hydropower. In Mozambique and Tanzania, vast quantities of gas have recently been discovered offshore, placing these two countries among the top six in SSA on P2 reserves (see table 1.1). There is also vast hydropower potential in Cameroon, the Democratic Republic of Congo, and Mozambique. The domestic energy resources in these eight countries indicate a high degree of firm energy,[1] from which power systems can benefit; this is especially important for power generation investments as such projects are more financially attractive and might result in lower unit costs. From a mining perspective, low-cost hydropower generation benefits extractive activities, but may attract primarily smelter activities, which require relatively cheap electricity.

The opportunities and scale of integration depend on many factors, including cost, electrification rate, and power reliability and adequacy. Reliability and adequacy of power from national utilities to large consumers is essential, particularly for mines requiring a continuous supply. Poor reliability presses mines to meet their power requirements through self-supply. Guinea's and Mauritania's utilities have low availability of already low installed capacity. Mozambique, on the other hand, is a stronger performer in both categories (figure 4.1).

Cost savings make a compelling case for power–mining integration. In Cameroon, the Democratic Republic of Congo, and Zambia, the grid tariff is less than 10 percent of self-supply, based on a typical 5-megawatt (MW)

Figure 4.1 Reliability and Resource Potential

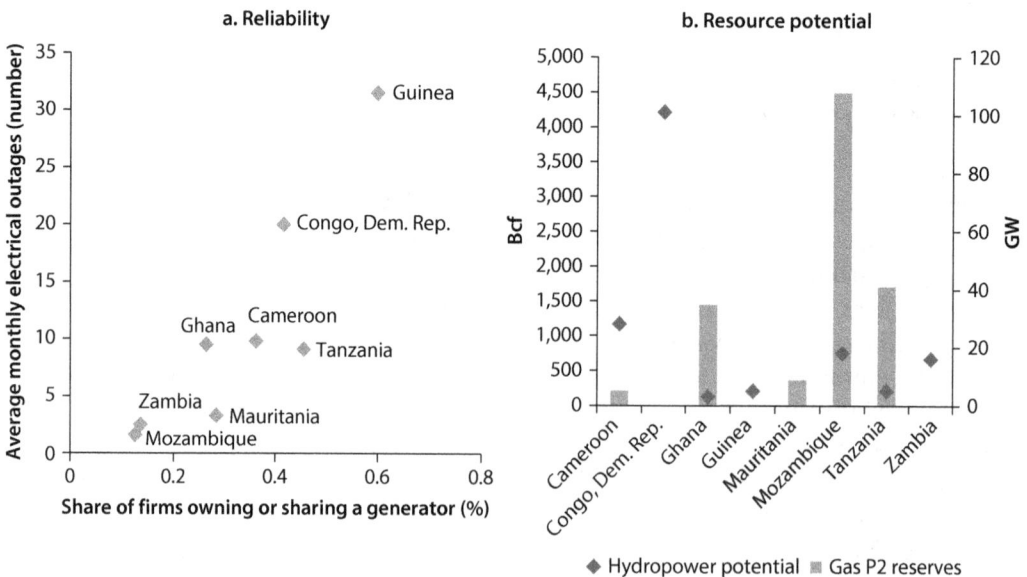

Sources: Economic Consulting Associates analysis; Foster and Steinbuks 2009.

Figure 4.2 Mining Tariffs and Power Sector Long-Run Marginal Costs

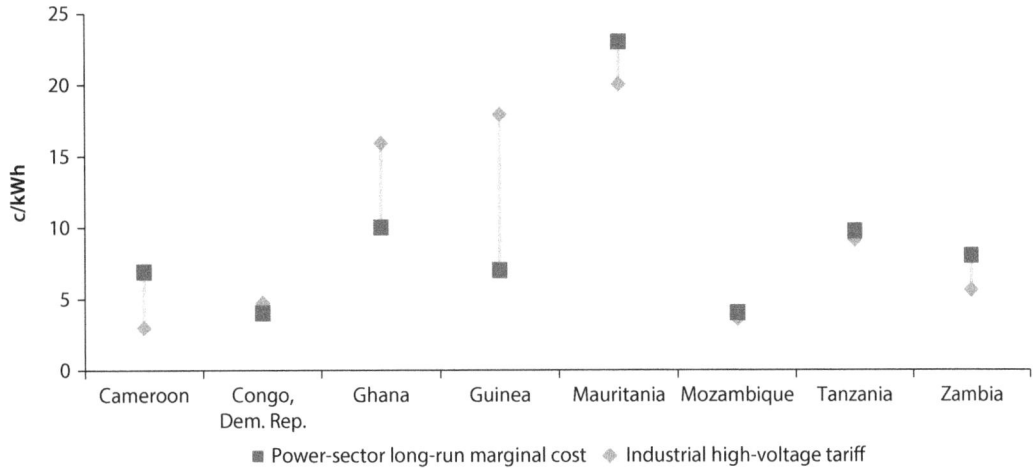

Legend: ■ Power-sector long-run marginal cost ◆ Industrial high-voltage tariff

Source: Economic Consulting Associates analysis.

diesel generator and compared with the system's long-run marginal cost (LRMC). Cameroon's mines pay the lowest tariff, followed by the Democratic Republic of Congo, Mozambique, and Zambia. In these countries, unreliability of supply must be considerable for the mines to switch to self-supply. The industrial high-voltage tariffs that apply to mines in Guinea and Mauritania are the highest among the eight countries studied (figure 4.2).

Key informant interviews with mining companies in the eight countries reveal a more complex set of objectives and motivations. Most important, mines are at least as concerned about supply security as they are about cost. They invest in self-supply to ensure continuous power availability even when the cost per kilowatt-hour (kWh) delivered will be much higher. Public hydropower stations with already written-down capital costs can provide particularly low-cost electricity, but self-generation becomes an option for mines if supplies of low-cost utility provided electricity are unreliable.

The Eight Countries Report a Range of Power-Sourcing Relationships

Mining power demand in 2020, including high- and low-probability projects, ranges from 125 MW in Mauritania to 1,394 MW in Zambia. As discussed in chapter 3, these eight countries represent a spectrum of mining power-sourcing arrangements, ranging from primary reliance on self-supply to complete integration with the grid. All of these countries have opportunities to use mining demand to benefit not only the mines but also the population and the country. More than half the incremental capacity over 2012–20 will be added to the intermediate category. About 24 percent of this new capacity will be grid-connected, while the remainder will use a self-supply arrangement. Only Zambia will stay

Figure 4.3 Power-Sourcing Arrangements (High- and Low-Probability Projects)

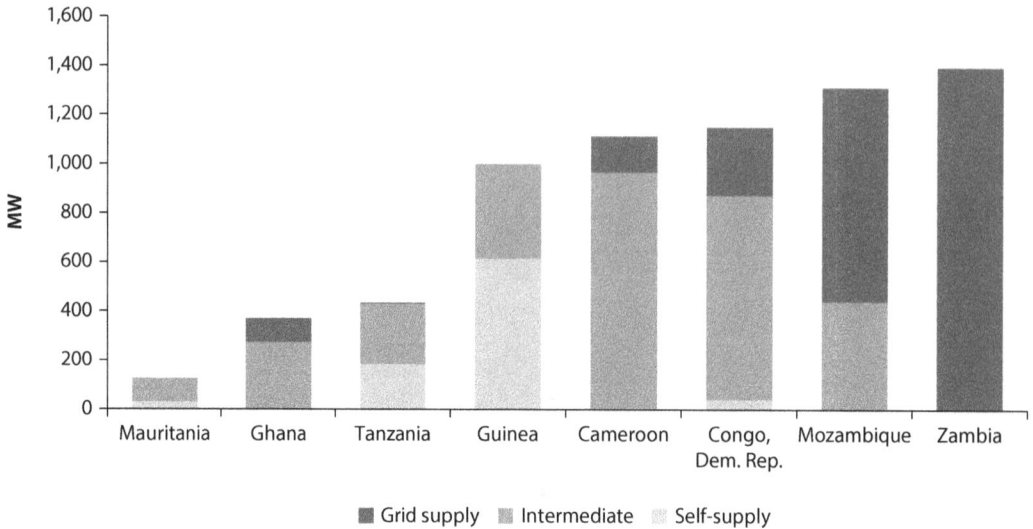

Source: Africa Power–Mining Database 2014, World Bank, Washington, DC.

completely grid-connected. Self-supply will be most important in Guinea, followed by Tanzania. Cameroon, the country with the largest low-probability demand, will depend almost entirely on intermediate arrangements (figure 4.3).

The eight countries were categorized into three groups, according to the synergies between their mining and power sectors (table 4.1). These groupings can change in accordance with the options' evolving economics.

Group 1: Minimal Synergies (Guinea, Mauritania, and Tanzania): Opportunity to Use Mining as Anchor Load for Electrification

Guinea

Guinea's electricity sector, which is vertically integrated and operated by Electricité de Guinée (EDG), faces major commercial and performance challenges. It has only 88 MW of operational installed capacity out of 395 MW total installed capacity and an electrification rate of just 18 percent. But the country is endowed with substantial hydro resources—an estimated potential of 6,000 MW. These resources have remained undeveloped due to a lack of financing for generation and transmission, insufficient residential demand to justify hydropower plant development, and nascent dialogue between the mining and power sectors.

Guinea has an active and growing mining sector, boosted by three key projects: Simandou (iron ore, Vale), Simandou (iron ore, Rio Tinto), and Kalia (iron ore) (figure 4.4). But the fate of the Simandou mine is uncertain. A sharp drop will occur in 2024, when the gold mine Kouroussa project is expected to end.

Table 4.1 Groupings of Power–Mining Synergies for the Eight Countries Analyzed

Level of integration	Country	Tariff[a]	Major energy resource of supply option	Utility reliability	Main ore production and mining process[b]	Major mining-supply arrangement
Minimal synergies	*Group 1: Minimal—Potential for mines as anchor demand for local development*					
	Guinea	High	HFO	Low	Bauxite, iron ore (future)	Self-supply
	Mauritania	High	Gas, HFO	Low	Iron, gold, copper	Self-supply
	Tanzania	Medium	Hydro, HFO	Medium-low	Gold, iron (future), nickel (future), uranium (future)	Grid (transitioning to self-supply)
Medium synergies	*Group 2: Medium—Potential for mines as anchor demand for national and regional development*					
	Cameroon	Low	Hydro	Medium	Aluminum smelting, iron ore (future)	Grid
	Congo, Dem. Rep.	Low	Hydro	Medium	Copper including smelting, coltan	Grid
	Mozambique	Medium	Hydro, coal	High	Aluminum smelting, coal, ilmenite	Grid or self-supply
High synergies	*Group 3: High—Potential for tariff rationalization*					
	Ghana	Medium	Hydro, HFO	High	Gold	Grid
	Zambia	Low	Hydro	High	Copper	Grid

Note: HFO = heavy fuel oil.

a. Low = 3–4 cents per kilowatt-hour (kWh); medium = 6–12 cents per kWh; high = > 13 cents per kWh.

b. These countries also have large energy resources.

61

Figure 4.4 Localized Simulation—Mining with Neighboring Residential and Nonresidential Demand, Guinea

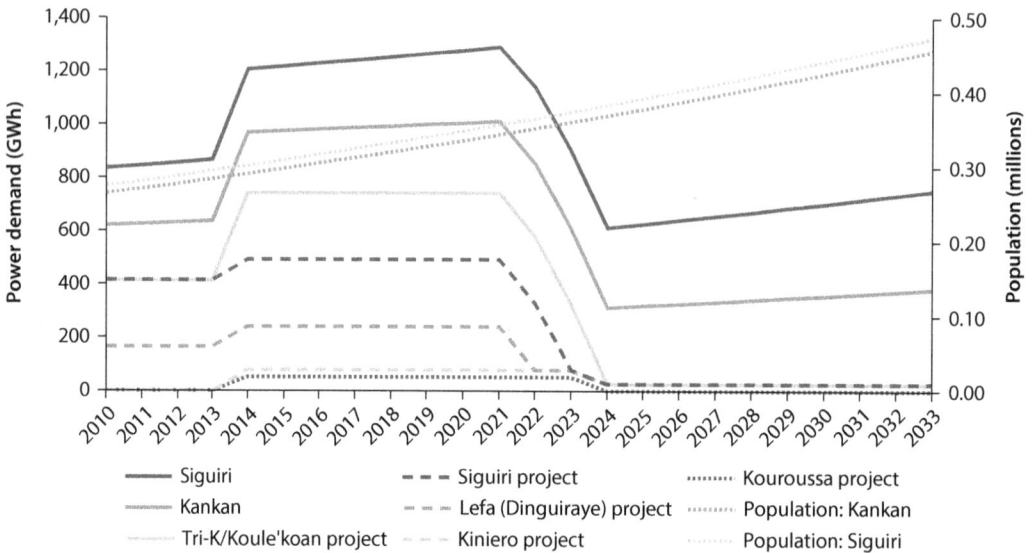

Sources: Africa Power–Mining Database 2014, World Bank, Washington, DC; Economic Consulting Associates analysis.
Note: The figure includes projects operational on or before 2020; demand in 2030 is captured by projects operational in 2020 that will still remain operational in 2030. A number of projects will close between these dates.

Demand from the iron ore mines becomes more dominant as the iron sector grows and as bauxite and gold mines exhaust their resources and close.

Self-supply is common among mining operations. An alternative scenario could be one or more isolated grids or a fully integrated national grid. But the situation presents a paradox: until there is a properly functioning system, the mines have no incentive to join; but without the large anchor demand that the mines can offer, old lines cannot be economically rehabilitated or new ones justified and financed.

The grid is nascent and will not soon extend to mining demand centers. Thus, the simulations elaborate on a location-specific analysis, where mining demand is aggregated in a group of five contiguous gold mines in northeast Guinea: Siguiri, Kiniero, Lefa, Tri-K, and Kouroussa. In addition to mining demand, the simulations compute residential and nonresidential demand in two neighboring towns, Siguiri and Kankan, based on each one's population, electrification rate, population growth, and typical household load. Nonresidential demand is assumed to be half of residential demand. As the towns grow, the rising dominance of their power demand coincides with a gradual depletion of mining resources. By 2024 most mines will have closed down, and the power system will operate to supply local towns and ore deposit sites next to the original mines.

The five contiguous mines have several available options, as follows:

Scenario 1—Mines self-supply. In this scenario, the mines continue sourcing their power from diesel-fueled self-supply units. Investment costs are relatively modest, but ongoing fuel costs and operating spending are high. Cost is only for generation, and developing transmission lines is unnecessary.

Scenario 2—Shared hydropower plant. A hydropower plant is developed close to the mines, and a mini-grid is built between the mines and plant. The plant acts as an independent power producer (IPP), with power purchase agreements (PPAs) from the mining companies as the viable off-taker. By this scenario, a 150-MW hydropower plant may be developed on Mandan River close to Kankan.

Scenario 3—Shared hydropower plant plus town supply. The IPP mini-grid is extended to the nearest towns of Siguiri and Kankan, with a total population of 539,306; the scenario assumes annual population growth at 2.4 percent, an electrification rate of 18 percent, and nonresidential demand at half of residential demand. The scenario involves expanding plant capacity to 300 MW and strengthening transmission lines that would connect the closest population.

For scenarios 2 and 3, it is assumed that the mines would jointly form, or otherwise contract, an independent IPP to manage the generation and transmission system in what would in effect be a mini-grid operating at transmission voltage. The investment in and commercial arrangements for the distribution network within the towns would have to be separately determined.

The results show that scenario 1—the status-quo option, where mines are autonomous and use decentralized diesel-fired generation—is the most costly outcome, with unit costs of 24.5 cents per kilowatt-hour (kWh) (figure 4.5). Scenario 2 has the lowest costs, at 4.9 cents per kWh; but scenario 3 is not much higher, at 5.0 cents per kWh. The key drivers of these collaborative cost savings are lower running costs and economies of scale from a relatively large hydropower plant. The addition of town demand in scenario 3 spreads the costs over a wider customer base and further drives down unit power costs for the mines.

Figure 4.5 Levelized Costs of Three Scenarios, Guinea

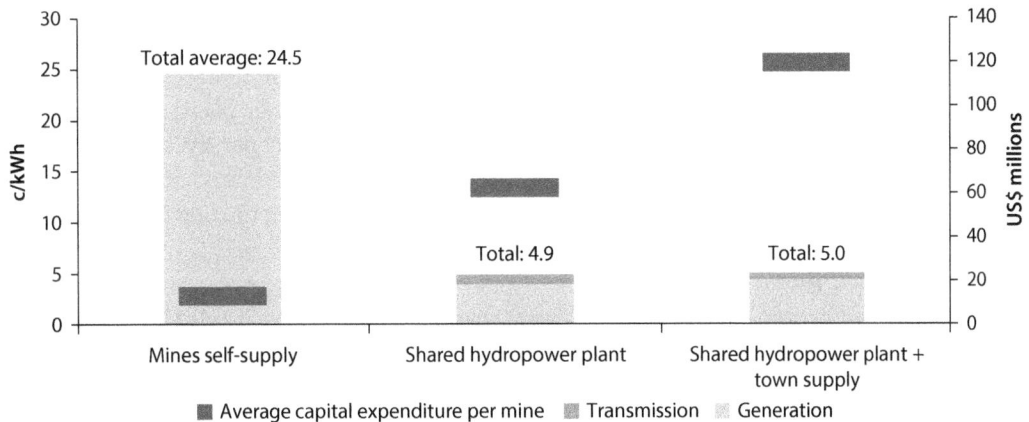

Source: Economic Consulting Associates analysis.

It appears unlikely that the mines would be willing to incur substantially higher capital costs to provide coverage for local communities when this would raise the levelized cost of electricity. To bring in the local communities, some form of government subsidy or development partner financing would likely be needed.

The capital cost of self-supply (scenario 1) is lowest, amounting to just $65 million. The hydropower and transmission options (scenarios 2 and 3) would require much higher capital investments of $310 million and $592 million, respectively. Not all capacity would be taken up by the demand included in scenarios 2 and 3, meaning that the hydropower plant could supply other customers in the future, particularly when mining demand falls as reserves are exhausted. Scenario 3 could reach 5 percent of the country's population. In scenario 1, carbon dioxide (CO_2) emissions total 2,831 tons, compared with no CO_2 emissions in scenarios 2 and 3.

The lower per-unit costs in scenarios 2 and 3 could lead to substantial economic benefits for both the mines and the population. If the mines adopted scenario 2 or 3 instead of scenario 1, they could gain about $640 million. The calculation for the population's economic benefit was based on per-unit cost estimates and the alternative cost of dry-cell batteries at $8 a month. Values depict benefits in current value terms over the project life. In scenario 3, the economic benefits in the electrified towns could total about $433 per household over the project life.

Results of a sensitivity analysis on fuel cost, discount rate, and capital costs show that scenario 1 is highly sensitive to fuel price (see "Generation cost," appendix C). If the diesel price rises by 5 percent, unit costs increase by 4 percent. Scenarios 2 and 3 are both sensitive to capital costs as the proposed hydropower plant is capital intensive. A variation of plus or minus 10 percent on capital spending will cause a corresponding per-unit increase or decrease of 5 percent. Results in scenario 3 also depend significantly on the underlying discount rate assumption: costs are frontloaded and not much affected by the discount rate, but a high discount rate makes for a higher unit cost in scenario 3, where town consumption extends beyond that of the mines.

A larger case study, including the Simandou mine and covering the entire eastern region, can also be simulated. However, because of Simandou's uncertainty, this case study is not elaborated here. In this case, electrification in neighboring communities can reach up to 10 percent of the population. The average generation cost for the mines that self-supply is 23.6 cents per kWh. The total unit cost for generation and transmission in scenario 2 (shared hydropower plant) and scenario 3 (shared hydropower plant plus town supply) is 6.2 cents per kWh and 5.5 cents per kWh, respectively. Obviously, scenario 3 results in the lowest unit power costs. However, the capital costs for scenario 3 are much higher, totaling $1,441 million.

Mauritania

The national utility, Société Mauritanienne d'Electricité (SOMELEC), is a state-owned, vertically integrated company responsible for generating, transmitting, distributing, and commercializing electricity in urban areas. Traditionally, mining in Mauritania has been dominated by the state-owned Société Nationale Industrielle et Minière (SNIM), but now the country also has many private companies. SNIM owns the majority of iron ore mines in the north, which constitute by far the country's largest mineral resource.

Mauritania's mines are located in remote areas without transmission grid systems. SOMELEC's limited power generation capacity has pressured the mines to secure their own power needs. To overcome these difficulties, the Mauritanian government is seeking to increase private sector participation in the electricity sector. A new power generation project with the private sector—the 80 MW Duale plant—is expected to switch from HFO to natural gas, with a total capacity of 300 MW, by 2016 when Banda gas becomes available. The dual-fuel power generation project, partially owned by SNIM and SOMELEC, will supply electricity to SOMELEC and export power to Senegal in the first phase. This plant has the capacity to produce additional power to be sold to a third party that could be a mining company. The transmission systems are nascent, apart from a high-voltage connection to Senegal.

Given the option of exporting 80 MW power to Senegal through the existing transmission line, the Duale plant becomes the marginal generation option until 2030 if Senegal is the off-taker or the mine off-takes 75 MW. If the mine off-takes only 25 MW, Duale is typically the marginal plant until 2035. Duale's ability to operate will be subject to both the level of off-take and load growth. From the utility's perspective, the plant will essentially be used to meet the country's load growth for the next 20–25 years. Not surprisingly, from an operational perspective on plant viability, there is little difference between Senegal off-taking 80 MW and mines off-taking 75 MW. However, if there is a financial requirement to invest in a transmission line between the plant and the mines, then exporting electricity to Senegal is a much preferred option as there is no need for a new line.

The simulation in Mauritania covers four mines in the northeast region (figure 4.6). Integrating all mining operations with the national grid is difficult because of the remote location of the largest iron ore mines. These mines can choose from the following three scenarios:

Scenario 1—Mines self-supply. The four mines continue meeting their power needs using diesel-fueled, self-supply units.

Scenario 2—Shared combined-cycle gas turbine (CCGT). A new 150-MW CCGT (E-type) gas-powered plant is developed in Nouakchott. The plant acts as an IPP with PPAs with the mining companies. The investment cost and grid maintenance are the IPP's responsibility, which may be owned by the mining companies or be a separate private entity.

Scenario 3—Shared CCGT plus town supply. This scenario expands on scenario 2 by extending the transmission lines to nearby towns.

In scenario 3, the plant can serve such neighboring towns as Benechab, Akjouj, Atar, Chinguetti, Aoujeft, and Zouerate, whose combined population of 143,436 is equivalent to 5 percent of Mauritania's total population and access deficit (see map A.3). This scenario assumes 2.2 percent population growth, 11 percent annual electricity growth, 19 percent electrification (as now), and nonresidential demand

Figure 4.6 Localized Simulation—Mining with Neighboring Residential and Nonresidential Demand, Mauritania

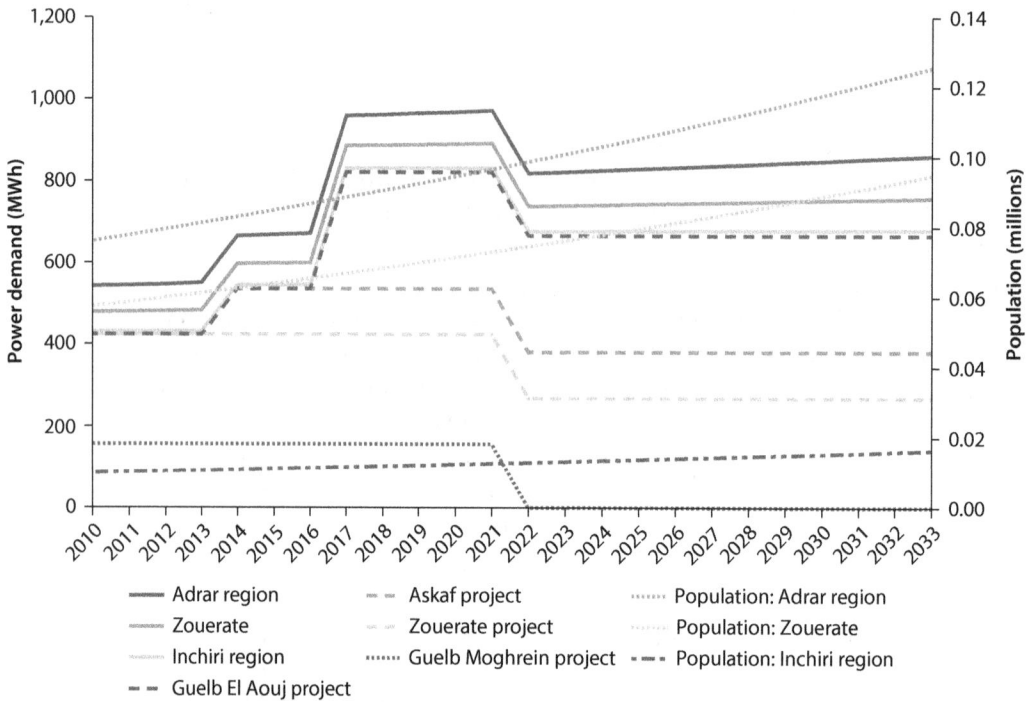

Source: Economic Consulting Associates analysis.

at half-residential demand. Estimates of residential and nonresidential demand for all regions total 126 gigawatt-hours (GWh) for 2013 and 183 GWh for 2030.

Another alternative could be a large-scale solar power plant in the northern regions. The gross solar energy input in Mauritania is estimated at 218 GWh a year. But this option must also be compensated with backup generation to support continuous mining activities. Solar cannot provide the capacity for machinery start-ups, so complementary diesel would be needed. Furthermore, solar power is intermittent. Only high-cost hybrid systems with backup generators and batteries could offer mines the necessary security of supply, yet with high risk in the isolated northern mining areas. But solar hybrid systems will become more relevant (Halley 2013). At this point, the most feasible option is developing the gas power plant in Nouakchott.

Mines would jointly form or otherwise contract an IPP to manage the generation and transmission system in a mini-grid operating at transmission voltage. The investment in and commercial arrangements for the distribution network within the towns would have to be separately determined.

Simulation results indicate that scenario 2 presents the lowest unit generation costs, at 9.1 cents per kWh, suggesting that the mines working jointly yield the highest economic benefits (figure 4.7). Further expansion can be considered as

Figure 4.7 Levelized Cost of Scenarios, Mauritania

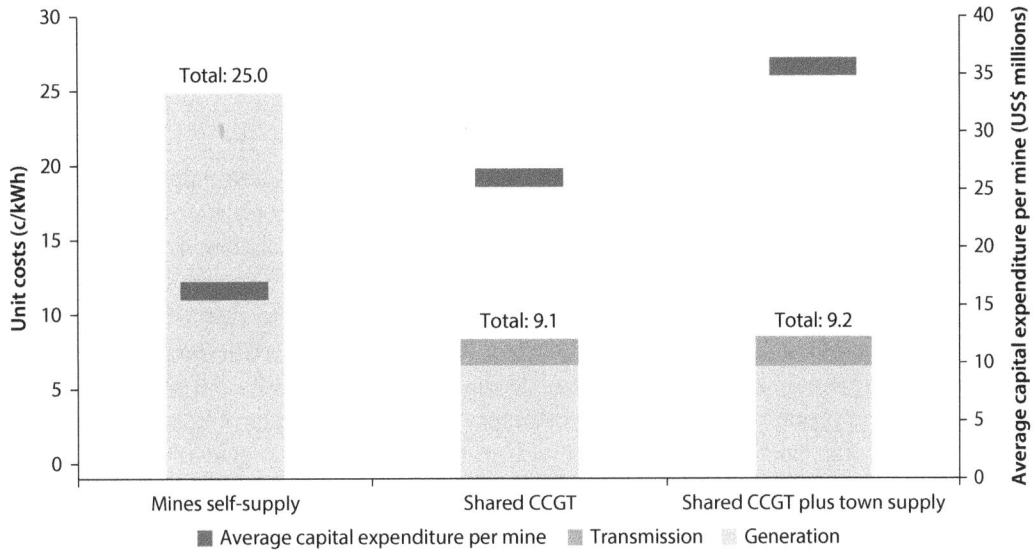

Source: Economic Consulting Associates analysis.
Note: CCGT = combined-cycle gas turbine.

the unit costs in scenario 3 are slightly higher than those in scenario 2. The most expensive solution would be continuing self-generation using diesel. The key drivers of the cost savings are economies of scale in investment costs and lower running costs of the CCGT plant, which offset the costs of the transmission lines. Urban demand in scenario 3 is not high enough to reduce the unit costs to compensate for the additional capital costs of connecting the town centers to the CCGT-supplied transmission grid (figure 4.7). To deliver electrification to the neighboring towns, external funding of the transmission interconnection may be necessary.

The total capital cost of the three scenarios is $60 million, $104 million, and $142 million, respectively. The project brings economic benefits for mines and the population. The mines stand to gain about $990 million, while the economic benefit is estimated at about $285 per household over 20 years.

CO_2 emissions are also important in choosing the optimum arrangement. Both alternative generation sources involve fuels with such emissions. But diesel's emission factor is much higher than that of gas, and the CCGT plant's efficiency is higher than that of diesel generators. It is thus expected that scenarios 2 and 3 will have lower emissions than scenario 1. Scenario 3 offers far higher electricity generation and consumption over the project life, so in absolute terms, at 5,737 tons of CO_2 (tCO_2), it has higher emissions than scenario 2, at 2,112 tCO_2 and even scenario 1, at 3,651 tCO_2.

The sensitivity analysis for the discount rate, cost of capital, and fuel costs suggests that the discount factor does not significantly affect the total per-unit costs (see appendix C). But 5 percent and 10 percent rises in fuel prices may raise unit costs by 3–11 percent. Scenario 2 is the least sensitive to fuel price variations, but both scenarios 2 and 3 are sensitive to capital expenditure variations.

Tanzania

Tanzania's electricity sector is managed and operated by the state-owned, vertically integrated national utility, Tanzania Electric Supply Company (TANESCO), which provides electricity to about 19 percent of the population. The country's main installed power generation capacities (1,006 MW) are based on hydropower (56 percent) and natural gas (34 percent). Despite the country's vast potential energy resources, its power sector has experienced several generation crises over the years due to low water level in hydropower dam reservoirs and delays in expanding generation capacities, leading to many blackouts and power rationing, and prompting large power customers to meet their needs outside TANESCO.

Tanzania is well endowed with mineral resources. Its gold reserves are considered second in Africa only to South Africa. It is one of a handful of countries where the mining power load is explicitly considered in the sector master plan. The plan's 2012 update provides details of new mining and other large incremental loads that TANESCO expects to go on the grid by 2018, for a total of 563 MW: 200 MW for two large projects—liquefied natural gas (LNG) and an iron ore smelter—and up to 197 MW from gold, nickel, and diamond mining.

The potential for linkages between power and mining emerges from TANESCO's poor financial position. The current tariff for the mines is 9.1 cents per kWh, which is close to the LRMC value of 9.7 cents per kWh. If TANESCO could fully satisfy mining demand, the annual power bill would be about $45 million. If high-growth conditions prevail, driven largely by growing mining demand, and if mining tariffs are set close to a rolling LRMC, then tariffs should decline to about 6 cents per kWh, a 34 percent reduction, toward the end of the planning period. Existing mines will save $15 million a year, and new mines will be more competitive on lower electricity costs. Lower operational costs will spur new mining investment, helping the country reach its ambitious long-term economic growth targets.

But the immediate problem is that TANESCO's poor service forces the mines to depend on backup generators for essential operations and cut back production to the extent that backup capacity is inadequate to allow full operation. Some of the mines have decided to invest in either diesel generators or diesel-solar hybrids to produce at full capacity, irrespective of whether power is available from TANESCO.

TANESCO's inadequate finances are a primary underlying cause of inadequate maintenance and investment, which lead to inadequate supply, load shedding, and frequent faults. As long as this vicious cycle persists, mines will depend more often on self-supply, incurring higher production costs and reducing output and thus tax and export revenues. TANESCO's inadequacies and associated costs have motivated mines to seek alternatives (table 4.2). The most important project is the 400-kilovolt (kV) backbone grid reinforcement project. Running from

Table 4.2 Cost of TANESCO's Inability to Supply the Mines, Tanzania

Party	Capacity (MW)	Annual electricity bill at full supply (US$ millions)	Cost of TANESCO inadequacy
Mines with modest standby capacity	45	28	Direct costs from standby generation; indirect (larger) costs from forgone production
Mines with adequate own generation	25	17	$50 million self-supply cost, an increment of $33 million
TANESCO	n.a.	n.a.	50+ MW to be supplied to other customers, with loss of net revenue

Source: Economic Consulting Associates analysis.
Note: n.a. = not applicable.

Iringa to Shinyanga (map A.4), it is crucial to ensure supply security for existing mines and make further mining development possible, especially for gold. Allied to this project, three generation and transmission project concepts have been developed that mines could collectively pursue to overcome power constraints while achieving much lower delivered power costs than diesel self-generation.

In Tanzania, simulations focus on the additional capacity requirements for new mining demand, connection of the new mines to the national grid, and efficient operation and maintenance of existing and new transmission lines and power plants.

Scenario 1—Mines self-supply. Mining demand will be satisfied from diesel-fueled, self-generation units. In this case, investment costs are relatively low, but ongoing fuel costs and operating expenditures are high.

Scenario 2—Shared power source for the mines. In this scenario, the power plant acts as an IPP with PPAs from the mining companies.

There are three options for this scenario:

- *Option 1—Hydro.* A hydropower generation plant is developed at Stieglers Gorge (Rufiji River Basin). This could create an opportunity for the country to develop the first phase of this project, as the mining demand could act as anchor demand and secure the investment. The first phase includes building a 130-m gravity arch dam and a 300-MW hydropower plant. A new transmission line would have to be built to connect the plant with the backbone 400-kV double-circuit transmission line in Iringa.

- *Option 2—Gas.* A 300-MW gas power plant in Dar es Salaam will be built to cover mining power needs after a new pipeline between Mtwara and Dar es Salaam is constructed. A new transmission line between Dar es Salaam and Dodoma is also needed to transfer the electricity to the backbone transmission line and ultimately the gold mines.

- *Option 3—Coal.* A 300-MW coal-fired power plant near the coal mines in Mbeya will be built to supply the northern region's gold mines. A new transmission line is necessary to take the power generated in Mbeya to the northern system.

Scenario 3—Shared power plant also serves town settlements in Mwanza and Shinyanga. In addition to electrifying the mines (scenario 2), the plant would feed into the national grid to meet the rising power demand in Shinyanga and Mwanza. The power plants' capacity would remain the same, except for the hydropower plant, where the third phase (the addition of a 600-MW unit) would also have to be developed.

Residential and nonresidential demand in the areas surrounding the mines is expected to grow rapidly. According to the national power system master plan, demand in Shinyanga is expected to grow from 391 GWh in 2013 to 3,778 GWh in 2030 and in Mwanza from 358 GWh in 2013 to 2,055 GWh in 2030. As a result, several additions to power generation capacity will be needed to meet rising demand projections.

The capital cost of self-supply is the lowest, amounting to an average of $50 million per mine. The options in scenarios 2 and 3 would require higher average capital investment levels. For example, the coal option would require $71 million for each mine, with totals of $300 million and $425 million in scenarios 2 and 3, respectively.

Among the options in scenario 2, option 1 has the lowest unit generation costs, but this result needs to be further analyzed, given the severe droughts in East Africa that have affected hydropower generation availability. Option 3 has the second lowest cost, and option 2 the highest. Scenario 3 spreads the costs over a larger demand base and so has the lowest unit costs. This suggests that supply from a common source, combined with feeding excess capacity to the grid, is the most efficient strategy (figure 4.8).

The lower per-unit costs in scenarios 2 and 3 could yield substantial economic benefits for the mines and the population. The calculation is based on the mines' per-unit cost estimates and demand and the population's per-unit cost estimates and willingness to pay. The values depict the benefits in present value terms over the project life. They indicate that the mines could gain from $3.4 billion to $3.7 billion if they adopt scenario 2 or 3 instead of scenario 1. The financial benefits over the project life are estimated at about $400 per household for the newly connected consumers.

Option 3 and diesel-fueled self-generation emit the most CO_2 emissions. Option 1 has zero emissions, while option 2 emits the least compared with alternative fuels for thermal generation.

The sensitivity analysis suggests that the discount rate in scenarios 2 and 3 may significantly affect total costs, not only mostly through the current value of electricity volumes but also through the current value of capital and operating spending.

Capital cost variations affect projects with high capital expenditure. The largest difference is for hydro projects, which vary between about –9.9 percent and 8.3 percent for scenario 1. Gas projects exhibit a variation of between –8.3 percent and 4.1 percent, and a coal project varies between about –4.5 percent and 4.1 percent for the base case. Scenario 1 is only slightly affected by differences in capital costs.

Figure 4.8 Levelized Cost of Scenarios, Tanzania

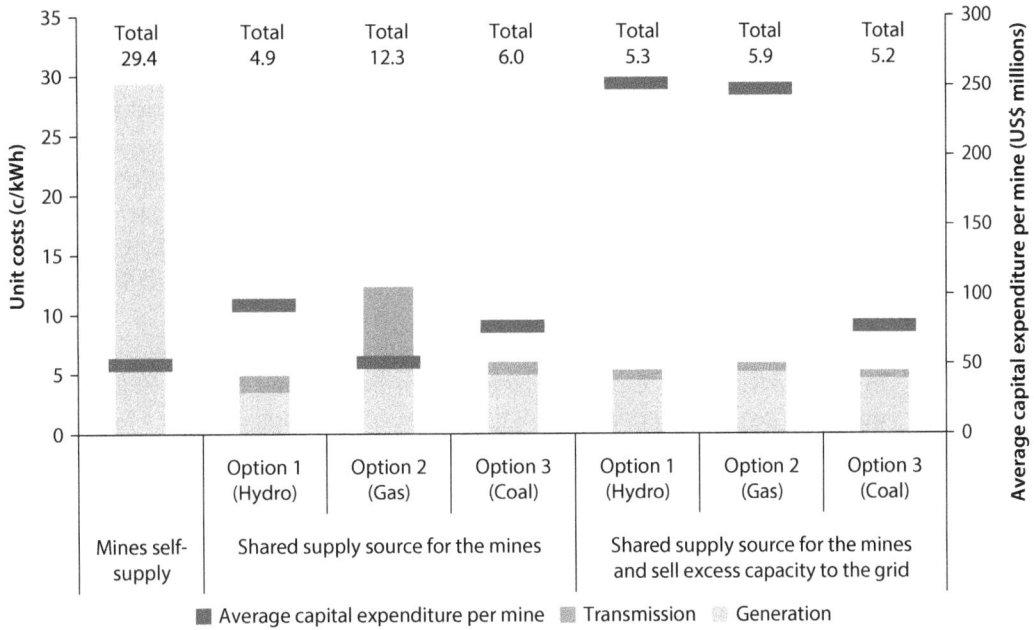

Source: Economic Consulting Associates analysis.

But scenario 1 is highly sensitive to changes in fuel costs, as of course total unit costs of a diesel-fueled generator are set mainly by spending on that fuel. A 5 percent rise in the fuel price causes a 4 percent increase in total unit costs; similarly, a 10 percent rise leads to a 9 percent increase. Scenarios 2 and 3 are less vulnerable to fuel-price variations.

Group 2: Medium Synergies (Mozambique, the Democratic Republic of Congo, and Cameroon): Mines as Anchor Load for Regional Power System Integration

Group 2 exhibits a higher degree of integration between the power and mining sectors due to the combination of large hydropower and gas resources and mineral deposits in these three countries. This generates investment alternatives in which the mining industry helps develop large power infrastructure projects important to meeting national and regional demands. These projects involve leveraging the mining companies' creditworthiness to finance power investments that would not otherwise be made, and they may not impose large costs.

Mozambique

Mozambique's electricity utility, Electricidade de Moçambique (EDM), is a vertically integrated company responsible for supplying electricity. It has an installed capacity of 2,428 MW and an electrification rate of 12 percent. The government has launched several large investment initiatives to harness the country's energy potential to meet its rapidly growing demand for electricity and export power to the Southern African Power Pool (SAPP) market, particularly to meet South Africa's heavy demand.

These initiatives consider the development of large power-generation projects, focused primarily on hydropower generation (1,500 MW, Mphanda Nkuwa, and 1,245 MW, Cahora Bassa North Bank), and the construction of high-voltage direct current and high-voltage alternating current transmission lines for evacuating relatively low-cost power (map A.5). A transmission system is to be owned and operated by a special-purpose vehicle, known as Sociedade Nacional de Transporte de Energía (STE).

Mozambique is well endowed with mineral resources; the coking coal reserves in Tete province are especially large and of exceptional quality. Starting from quite low coal production levels with its first coal exports in 2011, Mozambique is quickly becoming a major coal exporter; by 2025 it is expected to supply as much as one-quarter of the world's coking coal. In 2013–20, 190 MW of new mining demand is expected to emerge, driven by coal and iron ore.

The country's main load centers are distant from the hydro and coal complex in Tete. Transmission capacity is a key factor in realizing generation potential. The transmission studies associated with the generation master plan of 2009 estimated that more than $2 billion in investments would be required to evacuate power from the hydro and coal power stations. Financing these investments is by no means secured, and the shareholder structure is under discussion. But once the STE is built, different supply alternatives can be incorporated into the system.

The Mozambique case study differs from the others in that the main power–mining focus is not on the supply of electricity to the country's mines, but on the opportunity to generate low-cost power for the country and the SAPP market more broadly, through using discard coal from the coking coal export operations for power generation. Therefore, a separate option for creating a market for discard coal–fired power is simulated (figure 4.9), namely a smelter at the port of Macuze; this port is to be developed to handle the bulk coal exports from Tete.

Generating electricity from discard coal has obvious benefits, but with some offsetting costs. In the immediate future, mines avoid the costs of reburying the coal, secure their own electricity supplies, and earn a return on the power station investment through selling the power. EDM obtains a secure source of baseload power to supply the northern grid, with some grid reinforcement and local distribution, but there may also be grid stability issues. Mozambique avoids discard coal becoming an environmental hazard, but with the countervailing environmental cost of discharges from the burning of coal. If the regional market can be secured and the transmission capacity built so that discard coal generation can reach its full potential, the mines will benefit from increased turnover and the country will benefit through multiplier effects from the enormous boost to exports and GDP.

Figure 4.9 Demand of the Smelter and Coal Projects, Mozambique

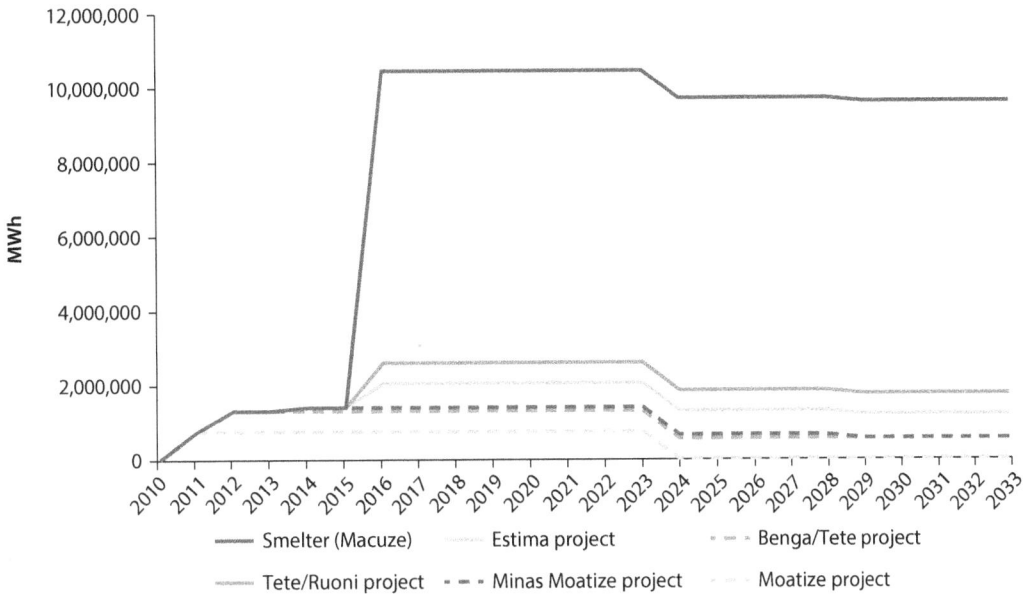

Source: Economic Consulting Associates analysis.

The two scenarios are as follows:

Scenario 1—Mines self-supply. Discard coal–fueled power plants at each of the mines meet only mine demand. This is merely for illustration; in practice, it would be expected that a certain amount of power would be exported to the national and regional markets (236 MW).

Scenario 2—Additional generation capacity, dedicated transmission line, and shared smelter at Macuze. Coal mines raise their power plants' capacity and send their surplus through a dedicated transmission line to a 900-MW aluminum smelter at the Macuze port (1,810 MW).

The average costs per kWh of power for the mines and the smelter project are 5.8 cents and 4.1 cents, respectively (figure 4.10). These costs include the capital costs of generation and transmission and an assumed coal price of $12.76 per ton. This a ballpark price that the mines might charge the power station if it is established as an IPP that also sells to EDM and SAPP markets. Because the opportunity cost of coal for the mines is close to zero, when they buy back power at 5.8 cents per kWh, they effectively pay only 4 cents per kWh. For the capacity created to supply the smelter, it is assumed that the discard coal is priced at zero, allowing power to be delivered to the smelter at 4.06 cents per kWh.

The CO_2 emissions in scenario 2 (the smelter) are far higher than in scenario 1.

Figure 4.10 Levelized Cost of Scenarios, Mozambique

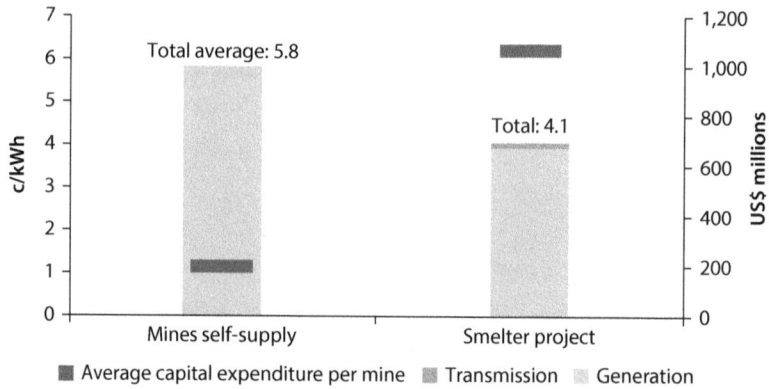

Source: Economic Consulting Associates analysis.

Democratic Republic of Congo

The Société Nationale d'Electricité (SNEL) is the Democratic Republic of Congo's government-owned, vertically integrated electric utility. Created in 1971, SNEL resulted from nationalizing several subregional private utilities to operate and maintain the Inga 1 and 2 projects and associated transmission lines. In late 2010 it was transformed into a limited liability company and placed under a performance contract signed with the state in early 2012. Despite SNEL's recent improvements in operational and financial performance, only 10 percent of households have access to electricity. Power outages averaging more than three hours in length are experienced more than 180 days a year. As a result, firms are forced to rely on expensive backup generators.

Since the mines have traditionally relied on SNEL grid supply, loss of production can be quite heavy when power supply is not secure. In February 2012 the Chamber of Mines estimated the regional power deficit in the mining-rich Katanga region at 141 MW, stating that, if it continued for a year, 250,000 tons of copper production would be lost (Munanga 2012). It would entail a reduction in turnover for the mines (at 2013 prices) of at least $1.8 billion; on an energy basis, it would equal about 170 cents per kWh or 36 times the official tariff for the electricity SNEL could not provide securely. The country overall would experience $1.8 billion in lost exports, 4.4 percent (some $700 million) in lost GDP, and lost tax revenues equal to 1.6 percent of GDP (about $250 million).

The mines are quite willing to pay higher tariffs to ensure more reliable supplies and end their production losses or quality penalties when variable power supplies interfere with production processes. A recent World Bank study on the Southern Africa Power Market project estimates that an overall tariff of 5.0 cents per kWh will be needed for SNEL, with a collection ratio of 95.8 percent (World Bank 2012). The corresponding high-voltage tariff of 5.5 cents per kWh allows for continued cross-subsidy of low-voltage customers, despite the higher unit costs imposed on the system. By 2019 the average SNEL tariff will need to rise to 6.0 cents per kWh.

The mining companies are willing to pursue relatively small hydro projects in Katanga province; the expectation is that once the Inga hydropower project is built, the country will be fully supplied with relatively cheap power, with an estimated long-run generation cost of 2.5 cents per kWh. From this perspective, any cost the mines incur through involvement in power projects or purchase of expensive power from local hydropower stations or through exports from Zambia is additional to what they would have paid had SNEL been efficient and invested enough to keep capacity at pace with demand at least-cost prices. In practice, that is not the case; the higher prices paid by the mines are much lower than the cost of forgone production.

Cameroon

The key issue for Cameroon is to develop its substantial hydropower potential, especially at sites along the Sanaga River. The completion of the Lom Pangar regulating dam will make it possible to generate up to 3,000 MW of reliable, all-season hydropower. Recognizing the potential of mining companies in developing hydropower, the government has taken a forward-looking, inclusive approach, involving the mining companies and other large electricity users in developing the new Electricity Act, which became law in December 2011 (see box 5.1). The deepening of power–mining relationships will yield substantial benefits for the country, the mining companies, the electricity supplier, and the currently dormant Central African Power Pool (CAPP). Mining companies will benefit from the large planned power projects and they are likely to contribute to the initial investment capital requirements given a clear regulatory framework. AES Sonel, the electricity supplier, will have additional capacity brought into the national power system without having to finance and to implement the hydropower and transmission projects itself. These projects will be large enough to fully exploit economies of scale, providing low-cost power to all consumers in Cameroon.

The CAPP region will benefit from expanded hydropower in Cameroon. Once local demand has been satisfied, it is expected that a surplus will be available for export. Cameroon is a key net exporter for the CAPP, and the Lom Pangar project is viewed as an essential first step for the country to realize its power export potential (Prevost 2010). The country will benefit from a substantial impact of projects of scale. The direct impacts will be through higher contributions to GDP, exports, tax revenues, and jobs, magnified by linkage and multiplier effects, not least the impact of improved supplies of least-cost electricity to the economy and the population.

Group 3: High Synergies (Ghana and Zambia): Lessons of Experience

Ghana and Zambia have long histories of engaging with mines in an integrated way. In Ghana, the giant Akosombo Dam was built on the Volta River in 1961–65 to power a 912-MW hydropower plant, which was upgraded in 2006 to 1,020 MW. Given that national demand in 1960 was only 70 MW, the decision to build the dam—with an aluminum smelter, Volta Aluminum Company Limited (VALCO), to serve as the anchor customer to develop the sector—was quite ambitious. Today the mining sector is relatively small compared with large industrial customers, with an estimated power demand of 95 MW in 2012–20. To date, the electricity sector is driven by business areas composed of generation under the Volta River Authority (VRA), transmission under the Ghana Grid Company, and distribution under the Northern Electricity Distribution Company and the Electricity Company of Ghana (ECG).

Zambia's mining-rich Copperbelt region has been integrally associated with developing the national utility, Zambia Electricity Supply Corporation (ZESCO). A state-owned company responsible for generation, transmission, and distribution, ZESCO supplies electricity to national and regional markets and covers about 23 percent of the population. Two other major players are the Copperbelt Energy Corporation (CEC) and the Lunsemfwa Hydro Power Company. CEC is a transmission company that purchases electricity from ZESCO at high voltage and distributes it to the mining industry in the Copperbelt region with some investments from the mining companies. CEC also has an 80-MW gas-turbine generation plant used only during mining emergencies. The Lunsemfwa Hydro Power Company is an IPP that generates about 48 MW of hydropower, which is sold to ZESCO under a PPA.

Mines in Ghana and Zambia rely primarily on the grid. But high dependence on hydropower and low reserve margins have made the power system in both countries vulnerable to periodic droughts. Ghana's recent capacity shortages, exacerbated by drought and the West African Gas Pipeline Company's gas supply problems, have resulted in load shedding and more frequent faults and outages. Although backup generators can help mitigate irregular grid supply, the direct costs to the mines are seen in lost production.

In Ghana, such imposed opportunity costs are quite high relative to normal power supply costs, and might justify installing enough diesel capacity to fully meet the electricity needs of large consumers, including the mines. An avoidable, major power crisis in 2006–07 is estimated to have cost the country nearly 1 percent in lost GDP growth (World Bank 2013). The current pipeline of national generation and transmission projects is sufficient to address the supply shortfalls experienced in recent years, so the mines no longer need to be proactive in directly securing their own power supplies.

In Ghana, tariffs negotiated by the mines are now higher than commercial and residential tariffs (low consumption). Historically, mining tariffs were lower. The mines did not have a financial incentive to sign PPAs with IPPs as they receive low-cost hydro or blended hydro from the generator (VRA). Typically, these agreements with the VRA were short, flexible, and revised annually, unlike the more stringent clauses in PPAs (capacity payments and take-or-pay) (World Bank 2013).

Given the uncertain timing of gas-to-power projects and the substantial rise in tariffs, the mines now have an incentive to develop power projects with IPPs.

The new tariffs, announced by the Public Unities Regulatory Commission (PURC) on October 1, 2013, require that large industrial consumers, including the mines, pay about 39 cents per kWh for additional power to meet demand. Industrial consumers with direct contractual relationships with the VRA reportedly have lower tariffs, at about 16 cents per kWh. Even so, these are quite high by historical standards as they are based on current thermal generation costs, with only 10 percent legacy hydro added to moderate the average tariff.

The legal framework associated with the proposed wholesale electricity market offers a way to overcome the lack of sector investment. A key regulation promulgated in 2008 (LI 1937) empowers the Energy Commission to register participants in the wholesale market, stipulating that the market will consist of two elements, a spot market and a bilateral contract market. Another clause in regulation LI 1937, yet to be implemented, prevents bilateral contracts, including legacy hydropower, to provide a strong incentive for the mines and IPPs to develop new generation projects to expand national capacity. Two major constraints must be overcome: lack of reliable and adequate gas supplies and lack of creditworthy off-takers beyond the mines due to poor technical and financial performance, particularly that of ECG.

In Zambia, the current mining tariff, equivalent to about 5.6 cents per kWh, is slightly lower than what residential consumers pay. So the question is, integration at what cost? New mining demand will stay grid-connected, and the sector has traditionally benefited from tariffs below what it should pay, based on a 2007 cost-of-service study (IPA Energy Consulting, Norton Rose, and PB Power 2007). The mines have no incentive to self-supply. But their contribution is limited to providing financing for connections to new mines remote from the grid, and new transmission lines and substations are needed.

The main commercial arrangement between the mines and ZESCO is brokered through the CEC. The relationship between ZESCO and the CEC is defined by a bulk-supply PPA, due to a lapse in 2020. According to IPA Energy Consulting, Norton Rose, and PB Power (2007), the tariff in the bulk-supply agreement is below the cost-of-service requirements (table 4.3); a more serious finding is that ZESCO has been pricing new contracts with the mines at rates that undercut the bulk supply agreement.

Table 4.3 Mining Tariffs and Cost of Supply, Zambia

Tariff (c/kWh)	2005	2006	2007	2008	Shortfall
Cost of supply for mining	—	2.65	2.76	3.01	—
ZESCO sales to CEC	2.27	2.33	2.41	2.48	0.53
ZESCO-Kansanshi PPA (2003)	1.80	1.85	1.91	1.97	1.04
ZESCO-Equinox PPA (2006)	—	—	—	2.14	0.87
CEC on-selling price	2.84	2.91	3.00	3.09	—

Source: IPA Energy Consulting, Norton Rose, and PB Power 2007.
Note: — = data not available; CEC = Copperbelt Energy Corporation; PPA = power purchase agreement; ZESCO = Zambia Electricity Supply Company.

Figure 4.11 Tariffs and Long-Run Marginal Cost, Ghana and Zambia

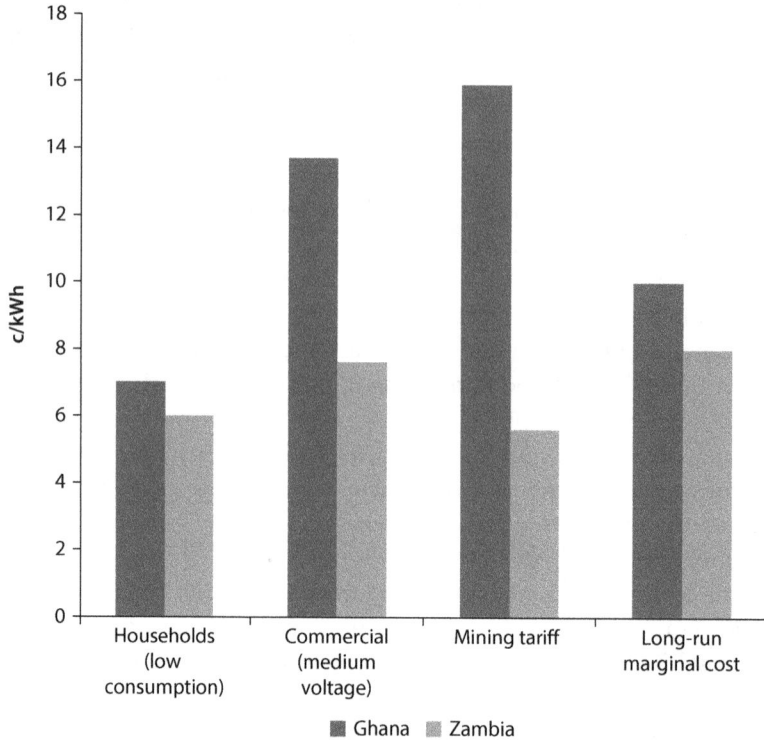

Source: Economic Consulting Associates analysis.

For a comparison of tariffs and LMRC in Ghana and Zambia, see figure 4.11. The 2007 cost-of-service study states that ZESCO's justification for exceptionally low tariffs in recent PPAs is that they reflect offset arrangements for investments made by the mines in ZESCO's transmission system for them to connect. However, such arrangements are not transparent. The study estimated that "if this underpricing of the mining load is allowed to continue, a $926 million deficit relative to ZESCO's Revenue Requirements will likely accumulate over the next 10 years." This would have to be funded either by cross-subsidies from other electricity consumers—potentially including low-income households—or from tax revenues, which would impose an opportunity cost on society at large, including the 80 percent of the population without access to electricity. Since 2007, mine tariffs have been raised twice; even so, they remain lower than commercial tariffs and the LRMC.

Note

1. Higher availability and less seasonal variability.

References

Foster, Vivien, and Jevgenijs Steinbuks. 2009. "Paying the Price for Unreliable Power Supplies: In-house Generation of Electricity by Firms in Africa." Policy Research Working Paper 4913, World Bank, Washington, DC.

Halley, Anthony. 2013. "The 'Revolution' in the Next Decade of Mining." *Mining.com*, March 12. http://www.mining.com/the-coming-revolution-in-the-next-decade-of-mining-88830/?utm_source=digest-en-au-130312&utm_medium=email&utm_campaign=digest.

IPA Energy Consulting, Norton Rose, and PB Power. 2007. *Revised ZESCO Cost of Service*. Edinburgh, Scotland: IPA Energy Consulting.

Munanga, Ben. 2012. *Impact de l'offre d'énergie électrique sur la production minière*. Unpublished manuscript.

Prevost, Yves Andre. 2010. *Harnessing Central Africa's Hydropower Potential*. Washington, DC: World Bank.

World Bank. 2012. *Africa—Second Additional Financing for Southern African Power Market Project: Restructuring*. Washington, DC. http://documents.worldbank.org/curated/en/2012/06/16360805/africa-second-additional-financing-southern-african-power-market-project-restructuring.

———. 2013. "Energizing Economic Growth in Ghana: Making the Power and Petroleum Sectors Rise to the Challenge." Washington, DC. http://issuu.com/fosu/docs/energizing_economic_growth_in_ghana.

CHAPTER 5

Challenges to Power–Mining Integration

This chapter traces the challenges of harnessing power–mining synergies, drawing on experiences from Sub-Saharan Africa (SSA) and elsewhere. It also considers risk mitigation instruments that will enable mining and power infrastructure to integrate, resulting in broader payoffs.

Three Scenarios

There are three scenarios for mineral-rich SSA countries (table 5.1).

In the first scenario, where there is no grid or the grid is too remote from the mining area and mines self-supply, self-supply options can expand to near-urban or rural locations. Mining companies, often located in remote areas, can help governments accelerate electrification of neighboring communities, thereby also raising the companies' social profile. Thus, many mining companies have embarked on electricity supply plans for communities, but they rarely fit them into a joint planning initiative that articulates company and government responsibilities—notably operating and maintenance, assessing the affordability to communities and their willingness to pay, and planning sustainable provision beyond the mine's life.

Mini-grid development could enable a more sustainable strategy for leveraging the mines' decentralized energy. Technologies using renewable energy sources have also become more popular among private participants, particularly mining companies. One recent example is the 2012 effort by Exxaro Resources, South Africa's second largest coal miner, to generate clean energy by establishing five renewable energy projects (two solar and three wind) in a joint venture with an undisclosed third party (Odendaal 2012). Another is the effort by Semafo Inc., a gold mining company in Burkina Faso, to build a 20-megawatt (MW) solar power plant under its subsidiary, Semafo Energy, in partnership with the Burkinabé government (Zgodzinski 2013). Mini-grids provide an excellent platform for instituting technologies for renewable energy sources.

Table 5.1 Power–Mining Integration Scenarios

Scenario	How can the power sector leverage mining energy demand?	Cost savings for the mine	Increased welfare for the host state
Grid: Too remote Mine: Builds its own generation ("Self-supply" and "Self-supply and CSR")	Leveraging decentralized energy for rural electrification (off grid or mini-grid)	Save the social license to operate	Accelerate effort of electrification
Grid: Too expensive or too unstable Mines: Builds its own generation ("Self-supply," "self-supply and sell to the grid," "mines sell collectively to the grid," and "mines serve as anchor demand for an IPP")	Leveraging for increased generation: If the mine produces excess and sells back to the grid If anchor demand for an IPP; if mines build bigger collective power plant	Either additional revenues Or diminished costs of energy needed	Additional sources of generation Cost of generation drops
Grid: Hydro-based (gas-based) and very cheap Mine: Wants to source from the grid ("Grid supply and self-supply backup," "mines sell collectively to the grid," "mines invest in the grid," and "grid supply")	Leveraging for more robust grid: If mines participate in upgrading the grid If mines leverage the idle capacity of emergency generators to alleviate the grid	Stable access to very cheap electricity Opportunity for additional revenues	Utility can gain in efficiency; infrastructure upgrading Avoid saturation of the grid

Note: CSR = corporate social responsibility; IPP = independent power producer.

In the second scenario, where grid-supplied electricity is more expensive or excessively unreliable compared with self-supply, mining companies have a clear incentive to invest in their own power generation and transmission capacity. The opportunity lies in the promise of additional generation sources, the challenge in developing incentives either for the mining industry to produce extra power capacity to be sold back to the grid or for mining demand to serve as an anchor demand for the independent power producers (IPPs)—both of which should lower the cost of electricity and increase the grid's stability. Given the capital spending in building self-supply and the scale economies to be gained in coordinated investments, there should be a business case for mines to coordinate joint investment. This only occurs with inexpensive sources of energy, such as hydropower and gas, and when a local partner takes part.

In the third scenario, where electricity provided by the utility is far less expensive than self-supplied power and relatively stable, mines will seek to buy electricity from the grid. The challenge is thus to find mechanisms to increase generation and transmission capacity and ensure unreliable supply. To continue accessing cheap electricity, mines will usually be willing to work with utilities and sometimes competitors under various commercial arrangements to set up or simply upgrade generation, transmission, and distribution capacity to meet their demand. The key is to find the right commercial framework that leads to cost savings for the mining industry and power infrastructure densification for the host country.

But the presence of moral hazard in this framework can lead to prioritization of mines over residential demand. To avoid this risk, supply- and demand-side management measures need to be executed, and regional power trade needs to be intensified. South Africa had a major power crisis in 2008, compromising the system to persistent shortages of about 3,500 MW (about 10 percent of peak demand) and crowding out residential demand. The government adopted a series of demand-side management and energy efficiency measures under its Power Conservative Program, with an initial focus on large industrial users, particularly mines and smelters. It saw impressive results within a few months (World Bank 2011b).

Constraints to Harnessing Power–Mining Synergies

Some barriers to power–mining integration are common across countries, while others are country-specific (table 5.2). In the Democratic Republic of Congo, Guinea, Mauritania, and Tanzania, constraints are closely related to low electricity tariffs, weak national power utilities, and inadequate national transmission grids. In Mozambique, transmission constraints are most pressing for coal-fired generation, both for supplying Electricidade de Moçambique (EDM) and accessing the

Table 5.2 Constraints to Power–Mining Integration

Constraint	Countries	Remedial policy actions
Inadequate national transmission grid	Cameroon; Congo, Dem. Rep.; Guinea; Mauritania; Mozambique; Tanzania; Zambia	Transmission reinforcement projects
Irregular fuel supplies and water flows	Cameroon, Ghana	Completion of Lom Pangar; backfeed to West African Gas Pipeline Company from Jubilee Field
Political economy and weak national utility	Congo, Dem. Rep.; Guinea; Mauritania; Tanzania	Utility and sector capacity building; strengthening regulators and their ability to raise tariffs to commercial viability levels
Rail and port infrastructure lacking for bulk mineral exports	Guinea, Mozambique	Rail and port projects
Regional market and interconnector capacity constraints	Congo, Dem. Rep.; Mozambique; Zambia	Reinforcement of regional market institutions and regional interconnectors
Investment frameworks	Cameroon, Ghana	Cameroon needs to firm up framework for mines to invest in hydropower projects; Ghana is poised to develop market rules to provide balancing market support for bilateral contracts between the mines and IPPs

Note: IPP = independent power producer.

regional Southern African Power Pool (SAPP) market. The Democratic Republic of Congo and Zambia must address regional power market and transmission interconnector constraints. Ghana's need to diversify from hydropower through gas generation is thwarted by unreliable gas supplies.[1] In Cameroon, an important constraint has been the control of flows on the key river with hydropower potential.[2] In Guinea and Mozambique, the expansion of iron ore mining and coking coal exports is constrained much more by transport than by electricity.[3]

The most frequent physical constraint is inadequate national transmission grids, incapable of handling additional flows as the mining sector and the rest of the economy expand. In Guinea, this lack is the most immediate, tangible obstacle. The state of the transmission network is largely due to Electricité de Guinée (EDG)'s long-term inadequacies. The physical infrastructure needs to be restored and strengthened as the utility is transformed. The localized power synergy proposal, elaborated on in chapter 4, could be a catalyst for this transformation. Mauritania has almost no transmission grid apart from a line connecting a hydropower facility on the Senegalese border with the capital city of Nouakchott, and this grid is neither strong enough for a sharp rise in demand nor—more important—located near existing or planned mining facilities.

The other dominant constraint is the utilities' weak financial situation. In the Democratic Republic of Congo, for instance, the government's comprehensive program for the financial and operational recovery of Société Nationale d'Electricité (SNEL) (2012–17) seeks to respond to the 2012 situation where high technical and nontechnical losses and low revenue collection rates meant that revenue was available from only slightly more than 1 kWh for each 2 kWh produced. Coupled with low tariffs, electricity sector inefficiencies were estimated to absorb as much as 4 percent of gross domestic product (GDP). In Tanzania, the Tanzania Electric Supply Company (TANESCO) has run financial deficits for many years; in 2012 its operating loss was $139 million, with a cumulative loss of $503 million (2 percent of GDP).

A Range of Factors Facilitate Successful Integration

Setting the Public Utility as a Viable and Creditworthy Partner for the Mining Company

Regardless of the power-sourcing arrangement between the mines and the public utility, the utility will be the mines' main partner. This is because, in all the countries except Ghana, a vertically integrated utility structure prevails, with 100 percent state ownership (except in Cameroon, where the American power company AES Corporation has a controlling interest in AES Sonel for the duration of the concession agreement). In Ghana, a separate transmission company has been formed (Ghana GRID Company), which acts as a single buyer from the generators and sells power to the distribution companies and export markets. Ghana has established the legal framework for a wholesale electricity market, but has yet to develop the regulations and instill the dynamism needed to get this market off the ground.

Whether the utility acts as a power off-taker, distributor, or co-investor, its financial health and creditworthiness will determine the range of possible power-sourcing arrangements. A stable utility is necessary for expanding the power sector and tapping into the potential generated by mining demand. Particularly important are transparent accountability mechanisms, dedicated investments to reduce inefficiencies and build capital stock, and a strong ministry of energy. For instance, some countries now use performance contracts, written agreements clarifying objectives and incentives to motivate managers to achieve them. They usually cover tariffs, investments, subsidies, social objectives, and funding, as well as performance indicators; sometimes they include rewards for good managerial performance and, less frequently, sanctions for nonperformance (World Bank 2011a). Other arrangements have similar objectives; for example, Cameroon has a concession agreement with performance indicators, and Ghana and Zambia set key performance indicators for utilities.

Integrating Power and Mining Master Plans

Among the eight countries, only two have explicitly incorporated power investment and mining growth into their power sector master plans (table 5.3). Mining investment thus cannot often be leveraged because the transmission network is not adapted to carry the load that mines could sell back to the grid.

Traditionally, the utility was responsible for planning and procuring new power infrastructure, but with power reform and introduction of IPPs, those functions have been transferred to the ministry of energy, which typically coordinates planning. In Kenya, regulators lead this task in an inclusive way. In countries such as Cameroon, Mauritania, and Tanzania, the ministries of energy and mining are integrated,[4] which could produce synergies.

Table 5.3 Power–Mining Coordination

Country	Mining demand explicitly included in sector master plan	Ministerial coordination between power and mining	Other coordination mechanisms between power and mining
Cameroon	No	Same ministry	Electricity Law of 2011
Congo, Dem. Rep.	No	Different ministries	Five-year program for financial and operational recovery of SNEL
Ghana	No	Different ministries	Ghana GRID Company wholesale power reliability program
Guinea	No	Different ministries	—
Mauritania	No	Same ministry	—
Mozambique	To some extent	Different ministries	—
Tanzania	Yes	Same ministry	The "Big Results Now" Program
Zambia	No	Different ministries	Office for the Promotion of Private-Power Investment

Note: — = not available; SNEL = Société Nationale d'Electricité.

Other coordination mechanisms exist. Through its Office for Promoting Private-Power Investment, Zambia is attempting to partner the private sector with the government to develop large hydropower and other generation projects. A best-practices example for other countries, Cameroon has adopted an approach that involves the mines and other large power users in planning how to exploit the country's hydropower resources and drafting the Electricity Law of 2011 to facilitate this (box 5.1). The framework requires that private developers compete for hydro sites, except when the site may more sensibly be allocated to a mine to develop power for its own needs. The challenges of ensuring that full economies of scale are realized and determining the optimal sharing of power between the mine and grid are neatly solved by requiring that the mine develop the full potential of the hydropower site and sell the surplus to the grid at cost-recovery tariffs determined by the regulator.

Mongolia provides an interesting example of an initiative to ensure multilevel coordination within government agencies and among government agencies, the mining industry, and civil society (box 5.2).

Introducing a Stable, Investment-Friendly Framework

Countries need to provide a sufficiently predictable environment to attract the massive investments warranted by Africa's rich mineral deposits. Most mining companies are keen to help develop gas or hydropower resources but will only do so if the country is stable, the policy risk is relatively low, and the legal framework provides clarity and certainty. Deficiencies in these aspects have deterred investors in Guinea and delayed hydropower expansion.

Box 5.1 Cameroon's Electricity Law

Cameroon's Electricity Law of December 2011 stipulates that the quantity of power sold back to the grid will, under terms of the concession agreement, be based on power optimization, and the tariff for electricity supplied to the public grid will be determined by the regulator on a cost-of-services basis to ensure that the general public also benefits from low prices. Sites must be allocated per competitive bid, but they can be awarded to integrated industrial projects on a sole-source basis.

In a policy letter signed by the prime minister, dated February 17, 2012, the government committed itself to develop all secondary legislation under the Electricity Law of 2011 in consultation with stakeholders. This secondary legislation will stipulate the principles to be used for determining the quantity of electricity allocated to the public grid, and it will include domestic supply-and-demand projections, the site's physical characteristics, the auto-producer's electricity demand, preference for supply to domestic ahead of industrial consumers or export, and existing arrangements between auto-producer's (self-supply) and the public grid concessionaire.

Source: World Bank 2012.

Box 5.2 Mongolia—An Institutional Framework for Planning

Mongolia's government has held negotiation forums with the mining industry to oversee infrastructure developments and define priorities, with plans to create new institutions to further this goal. One body would be granted the right to lead on infrastructure development decisions and implement the overall integrated development plan that each agency, according to its particular specialization, would be responsible for implementing, as follows:

Southern Mongolia Infrastructure Council. The council would comprise representatives from the national government, local governments, mining companies, and civil society. Its goal would be to serve as a forum for public consultation and information exchange. It would be developed as either an advisory committee or an entity that makes decisions and finances infrastructure development.

Southern Mongolia Infrastructure Coordination Unit. This entity would serve as an information forum to coordinate the multiple levels of local government, and would be entitled to step in to accelerate decision making.

PPP Unit. This unit would have expertise in public–private partnership (PPP) developments to compensate for the country's lack of expertise in this field.

Risk Management Unit. Since investors in PPP transactions typically request government guarantees, this unit would specialize in negotiating government guarantees for PPPs, and would set caps on governmental risk exposure. The unit would report annually to the government on the extent and probability of its liabilities.

International Infrastructure Expert Advisory Panel. This panel would ensure that the government negotiates the best deals, and might call on a panel of international experts to review cases.

Economic Regulation Agency. This agency would have expertise in tariff setting for rail and electricity sectors.

Southern Mongolia Groundwater Management and Information Center. This agency would be charged with gathering groundwater information from all other government agencies (World Bank 2009).

Source: Adapted from Vale Columbia Center, Columbia University 2012.

Public-private partnerships (PPPs) are allowed in all countries except the Democratic Republic of Congo, which is close to promulgating a law. Their legal framework ensures a formal articulation of asset status and ownership, dispute resolution mechanisms, service and performance obligations, and responsibility sharing between private and public parties, including when the partnership is terminated.

Regulation by contract is common practice in countries with a recent track record of PPPs, but it will only work with a supportive legal environment that enforces contracts and provides a clear and nonconflicting framework. It can also generate challenges and risks for host countries. Success depends on the government's capacity to negotiate a very complex contract and honor its commitment to transparency, given that the contract is bilaterally negotiated behind closed doors.

Where expanding generation capacity is required from the mines acting as IPPs, the investment costs must be recoverable and revenue streams sufficiently definite into the future to enable the power generation owner to obtain financing on reasonable terms. Regulations may allow providers and customers to enter into long-term contracts, whereby the customers (that is, the utility or other users) commit to buying a minimum capacity from the owners over a longer period. This is generally preferred by the providers as it is more certain and is usually necessary to obtain financing for investment. But the utility can be put in a difficult financing position when it is a contracting party, thus necessitating an appropriate framework for such contracts (usually power purchase agreements [PPAs]).

PPAs can be signed between the mining company and the utility (most commonly), between the mining company and an end user (when authorized), and between the mining company and an IPP (when authorized). Mining companies could be interested in signing with an IPP when it provides cheaper electricity than the public provider, when the public provider does not provide enough electricity, or when the mining company decides that electricity generation is not part of its business. Under this model, the IPP could either bear the risks and obligations associated with ownership, including commercial risks and maintenance obligations, or just be the operator of the power plant that the mine would finance. A case example from South Africa elaborates on a process followed, along with the roles of various stakeholders (box 5.3). But for the mines, IPP, and utility to come together, a more standardized arrangement that reduces the transaction costs for all parties is required.

Enacting Effective Regulations

Effective regulations on electricity provision are fundamental to ensure that the power and mining nexus is mutually beneficial and supports service expansion. Therefore, transparent and efficient regulations are required for electricity pricing, transmission services to or by the mines, and concessions for distribution or retail services provision by the mines. The tariffs should be transparent to all customers and reflect the cost of delivery or "usage of the system." Additionally, it is fundamental to disassociate the cost of a possible public-private investment from electricity pricing. This will allow transparent cost allocation of the investment and prevent temporary tariff privileges that would be difficult to remove.

Among the eight countries, Zambia has the longest established regulator (Energy Regulatory Board). Cameroon, Mauritania, and Mozambique also have energy or electricity regulators, while Ghana and Tanzania have multisector regulators (water and energy); however, Ghana's licensing and technical regulations fall under a separate body (Energy Commission). In the Democratic Republic of Congo and Guinea, which lack stand-alone agencies, the government is the regulator.

In most of these countries, technical regulation is satisfactory, but economic regulation appears weak in precisely the realm that is most critical—setting

Box 5.3 South Africa—Anglo American and Its IPP

To power its platinum mine, which requires a secure power supply for continuous operations and future expansion, Anglo American is seeking to sign a 450-megawatt (MW) coal-fired power project with an independent power producer (IPP) in Emalahleni municipality, South Africa. It is a build-operate-own project that plans to start commercial operations in 2015. The project "navigate[s] the relatively uncharted territory related to third-party use of the transmission and distribution system" (T. Creamer 2011). Tasks are split among the parties as follows:

Anglo American. The company provides the land, coal, and water (M. Creamer 2011). Anglo American is also in charge of securing financing through international loans; capital spending is $1 billion (Hall 2010).

IPP. Anglo American plans to sign a coal-supply agreement and a 25-year power purchase agreement (PPA) with the IPP to buy its entire capacity. It will also sign supplementary supply agreements with Eskom (the public utility) to use the electricity produced at the IPP plant for mining operations (Hall 2010). In parallel, the IPP signed connection, transmission, use-of-system, and operating agreements with Eskom to allow the IPP to sell its electricity to Eskom. An agreement was also signed between Anglo American and Eskom for Anglo American to off-take power from a substation to be built by Eskom (M. Creamer 2011).

Public electricity provider. Under the government's electricity strategy, laid out in its Integrated Resource Plan, Eskom is in charge of determining the terms of the connection agreements, infrastructure timing, and system cost use (Department of Energy 2011). (Anglo American criticizes this legal framework, judging that costs are too high by international standards, and that there is insufficient support to guarantee a fair allocation of costs.)

Role of other public agencies. The National Energy Regulator of South Africa and Department of Energy facilitate the contractual arrangements for third-party use (regulatory framework, appropriate pricing, timing of connection), and they must approve them.

Source: Adapted from Vale Columbia Center, Columbia University 2012.

cost-recovery tariffs that will make the utilities viable enough to properly maintain existing equipment and make investments. Two of the regulators with good governance credentials—Zambia's Energy Regulatory Board and Tanzania's Energy and Water Utilities Regulatory Authority—have nonetheless failed to raise tariffs to sustainable levels.

Regardless of the power market structure, regulatory oversight of the tariffs charged by the mining company selling under a PPA is needed to ensure a viable market for end users. The cost of bulk power supply is 50–70 percent of the distributor's total supply costs (Besant-Jones, Tenenbaum, and Tallapragada 2008). In SSA the price charged by the 28 IPPs under a PPA is 4–40 cents per kWh, with the upper bound often unaffordable to public utilities (Besant-Jones, Tenenbaum, and Tallapragada 2008). But to motivate companies to generate extra electricity, prices cannot be set too low. A solution could be to have a light-touch system, whereby the regulator does not set the prices but reviews the prices fixed by the parties and issues comments on their reasonableness until

Table 5.4 Regulatory Mechanisms

Country	Regulator (agency or contract)	Grid code
Cameroon	ARSEL	Yes
Congo, Dem. Rep.	Ministry of Energy	No
Ghana	EC and PURC	Yes
Guinea	Ministry of Energy	No
Mauritania	Regulatory authority	No
Mozambique	CNELEC	No
Tanzania	EWURA	Yes
Zambia	ERB	Yes

they reach an adequate level, as done in Nigeria (Nigerian Electricity Regulatory Commission 2006).

An effective regulatory system is also needed for managing risks and regulating access. Risks with IPPs and PPAs involve defaults, payment delays, and failure to honor contract obligations. Regulators can enforce contracts and strengthen the position of utilities that cannot provide sovereign guarantees, such as through escrow accounts, repatriation of profits, and guarantees against nationalization (required when the utility is not viable).

Regulators also need to ensure access. When mining companies as IPPs are authorized to sell to a third party, their access to the transmission network at nondiscriminatory tariffs must be ensured, particularly when the utility might be tempted to raise its prices for competitors and favor the electricity produced by its own generators. Eskom in South Africa has been criticized by mining companies for exercising this type of discrimination (Harding 2012). To avoid it, a grid code promoting open access at nondiscriminatory tariffs with associated dispute resolution mechanisms is necessary, but half of the eight countries lack one (table 5.4).

Risk Mitigation Instruments Exist

Carefully Drafted Contracts—Streamlined Concession Agreements for Retailing Electricity Services in Exclusive Areas Managed by the Mine

Where there is no national grid, as in Guinea or Mauritania, location-specific projects can be proposed. For new mines and surrounding communities, the mines would jointly form or otherwise contract an IPP to manage the generation and transmission system in what would, in effect, be a mini-grid operating at transmission voltage. The investment and commercial arrangements for the distribution network within towns would have to be separately determined. The national utility may choose to install and operate the distribution system within the towns, buying the electricity in bulk from the mini-grid. Or this could be done by a private distribution company or the IPP that operates the mini-grid. In the latter case, a certain distance between transmission and distribution would still be desirable to ensure a clear price at which the mini-grid delivers bulk electricity to the distribution entity, for retail sale to customers in towns.

Carefully drafted contracts are important. Some mining companies insist on providing local electricity as part of their corporate social responsibility (CSR) policy, but this initiative's voluntary nature often undermines its sustainability and absolves national and local governments of any responsibility. Developing model concession agreements mandating electricity provision within a certain radius would increase certainty for investors, put all mining companies' CSR programs on an equal footing, and enhance the government's accountability as the authority to enforce the contract.

If the scenarios are pursued further, it is necessary to spell out commercial arrangements for town electricity supplies, estimate the costs, calculate the tariffs, and then check whether the original demand assumptions still hold. The retail tariff may need to be subsidized; this could be done by the government through the national utility or direct payments, and the mines could also contribute, either voluntarily or through regulatory fiat. Mines would still have room to benefit from lower electricity tariffs relative to their next best option, while paying a premium on the calculated cost-recovery level; the additional revenue would go toward reducing the retail tariff in the towns.

Zambia, which has a distribution and supply company (Northwest Energy Company [NWEC]) already providing electricity to remote communities living near mines, offers a best-practices example (box 5.4). NWEC buys electricity in bulk from the national utility, Zambia Electricity Supply Corporation (ZESCO), and sells to some 1,000 retail customers that reside near the Lumwana mine.

Standard Commercial Arrangements

For mines taking supply at transmission voltages from the national grid, the standard commercial arrangement is a tariff appropriate for that voltage level, which would thus have both capacity and energy components. Where mines have paid some of the investment costs for the connection (transmission line and substations), it is common for the infrastructure to belong to the national utility and the prepayment to be treated as a loan (box 5.5). This is repaid in kind, rather than in cash, through an offset arrangement in the invoicing for power purchased by the mine. This could be made equivalent to a discounted tariff applied during the repayment period, but a discounted tariff can readily be misunderstood by other electricity consumers as discrimination favoring mining companies (an issue particularly in Zambia).

Given the perspective of savings when the grid energy source is hydro or thermal, mines are eager to connect to the grid and coordinate with and extend funding to the utility.

Box 5.4 Franchisee Experience of Serving Communities in Zambia

In Zambia, the Northwest Energy Company (NWEC) and the Energy Regulatory Board (ERB) recently had a tariff dispute. ERB asserted that the 9.5 cents per kilowatt-hour (kWh) tariff had not been approved, and should be reduced to the uniform national level of 6 cents per kWh with damages paid. In 2011 ERB took the dispute to the high court, but the judge dismissed the case. While 9.5 cents per kWh may seem high in Zambia, it would not be considered high in other parts of southern Africa, particularly in remote centers that would not otherwise have electricity supplies.

NWEC customers did not view the tariff of 9.5 cents per kWh as problematic, as evidenced by continued growth of the customer base. Furthermore, other mining companies have approached NWEC about setting up similar distribution mini-grids for First Quantum Minerals to electrify about 5,000 houses to be built by Kansanshi Copper and Gold Mines in Solwezi. And another larger project is on offer in Kalumbila. Had ERB won its case in the high court, NWEC might well have gone out of business, depriving the electricity consumers not just in Lumwana but in the new project areas offering power access.

This experience offers lessons for other countries seeking private operators of small distribution systems. The arrangement is likely to work only if the distribution operator is allowed to charge tariffs higher than the utility-supplied tariffs in the country's main urban centers. The arrangement can still benefit the customers served since electricity access will allow them to enjoy higher quality energy at a lower price than they would otherwise pay for inferior substitutes (such as kerosene or dry-cell batteries).

Box 5.5 Scope of Commercial Arrangements between Utility and Mines

Mines invest in transmission lines and the utility repays with interest. In Burkina Faso, Semafo, the owner of an open-pit mine inaugurated in mid-2008, Mana, signed an agreement with the government in October 2011. The commitment was to build a transmission line to Semafo's Mana mine at an estimated cost of $19 million. The line would reduce the mine's costs by $40 per ounce. Sonabel, the national power utility, would receive half the money from Semafo. Sonabel would repay it over eight years following commissioning. This investment would lower energy costs for the mine from 31 to 18 cents per kWh (Hill 2011).

Mines invest in power infrastructure and are compensated by reduced or negligible utility bills. To avoid costly self-supply, Katanga Mining Limited in the Democratic Republic of Congo took over upgrading the national grid. In March 2012 it announced that it had signed an agreement with Société Nationale d'Electricité (SNEL), the public utility, for a $283.5 million loan to upgrade the country's electricity generation and transmission networks. The company will be reimbursed $189 million by its affiliates at the mines of Kansuki and Mutanda, which will use much of the new electricity produced. Under the agreement, 10 percent of the power generated will be excess and sold back to SNEL. Of this investment, $261.8 million will be reimbursed through utility bill credits, and SNEL will also pay interest on the loan. According to

box continues next page

Box 5.5 Scope of Commercial Arrangements between Utility and Mines (*continued*)

Katanga Mining Limited, the new 450 megawatts (MW) of capacity—to be reached by the end of 2015—will enable copper production of 310,000 tons a year (Katanga Mining Limited 2012).

Mines invest in emergency power plant and get priority access during load shedding. In Ghana, as a result of the 2006–07 energy crisis, an 80-MW dual-fuel thermal plant at Tema was set up by a consortium of four mining companies. Completed in 2007, a part of the agreed ownership was transferred to the Volta River Authority (VRA) (GBC News 2008). Today, the plant serves the mines only as a backup against another energy crisis.

Mines pay back their priority access to power with a premium rate or even assistance in debt repayment. Zimbabwe's electricity relies largely on imports from Hydropower plant Cahora Bassa (HCB) of Mozambique. But in 2012 HCB threatened to cut power to the public utility, the Zimbabwe Electricity Supply Authority (ZESA), until it paid HCB its debt. Helped by two platinum-mining companies, ZESA cleared its $76 million debt with HCB by the end of 2012. It obtained $40 million from them, and credited the amount to their accounts as prepayment for electricity (Sibanda 2012). Also, ZESA guaranteed power supplies to the companies' operations for five years (Moyo 2012).

In Zimbabwe, New Dawn Mining Corporation's Turk-Angelus gold mine, located 56 kilometers (km) northeast of Bulawayo, is expanding to 23,000 ounces of annual gold production. Connected to the national power grid through an 88 kilovolt (kV)-ampere line, the mine has three generators used as a standby during any faults that can supply 3 MW of power. The generators were commissioned in November 2010, the same time that ZESA announced the introduction of an uninterrupted electrical supply arrangement with power charged at a premium rate. Given that the premium rate was still lower than the cost of operating the generators and that a suitable power line was available, the mine formed an agreement with ZESA. The generators will provide backup in case of regional or national power cuts. Once the uninterrupted supply proves reliable, the generators may be moved elsewhere (Othitis 2012).

Mines pay higher tariffs in exchange for utility investment in generation and transmission. In March 2012 the Zambian independent power transmission group, Copperbelt Energy Corporation (CEC)—owned by the mining companies and supplying power bought from the Zambia Electricity Supply Corporation (ZESCO) to most mining operations in Zambia's Copperbelt—and Nigeria's Africa Finance Corporation signed a deal to finance the construction of two hydropower projects in Zambia: one at Kabompo gorge in the northwest and the other in Luapula province. At a cost of $1.15 billion, these projects can provide a power surplus of 6,000 MW by 2016 (Chanda 2012).

The expansion of Zambia's copper industry has been hobbled by an unreliable power supply. The CEC warns that industrial tariffs will need to rise by 20–30 percent a year to reach cost recovery and support new investments in generation (CEC 2012). In 2008–11, mines were protected by a tariff stabilization agreement. After the agreement ended and under regulatory approval, ZESCO increased its bulk supply tariff to CEC by 30 percent. As a result, ZESCO and CEC must negotiate a five-year framework for further increases to reach cost-reflective levels for bulk supply tariffs (ERB 2011).

Source: Adapted from Vale Columbia Center, Columbia University 2012.

Mine-Financed Power Projects

Mine-financed power projects in association with the public sector need to follow transparent and effective PPP agreements, preferably detached from energy pricing. Mixing participation (through financing or capital lines) of mines in power infrastructure with tariff discounts will always be difficult to regulate. Electricity purchase offset arrangements are the norm for mines making substantial power investments beyond their mine connections; these investments are made with or on behalf of the national utility. Among the eight countries, an example is found in the Democratic Republic of Congo.

In the Democratic Republic of Congo, large projects financed by mining companies, which went ahead under arrangements with the utility (SNEL), proved unsatisfactory. After four years of negotiations, the $320 million Katanga Mines project began under a joint-venture structure. But when the project failed to move ahead, a turnkey contractor was appointed, jointly supervised by the mining company and SNEL. Also in the Democratic Republic of Congo, the $170 million Freeport McMorans project at N'seke is being implemented by SNEL. Even though the mining company is a member of a joint technical committee, it could not prevent the project from incurring substantial time and cost overruns. The mining companies should have chosen alternative investment structures for these projects, possibly opting for direct, jointly supervised turnkey arrangements (box 5.6).

Managing Mining Revenue for Public Infrastructure Services

Revenues generated by the mining sector via taxes, dividends, and/or royalties offer the potential to serve as an important driver of infrastructure development, if properly managed. Securitization can serve to improve borrowing conditions to investment grade and reduce financing cost; mitigate revenue uncertainty via risk sharing; promote strong governance, transparency, and solid asset management practices; and develop a capital market track record benefitting future transactions.

Box 5.6 Typology of Financing Arrangements

Investment in infrastructure that benefits from mining revenues can be separated into three classifications: the expansion of infrastructure services directly related to the mining industry to serve the wider public, financing infrastructure services indirectly related to the mining industry supported by mining revenues, and capitalizing mining revenue to fund infrastructure that is unrelated to the mining industry.

Expanding infrastructure services directly related to the mining industry to serve the wider public. Oversizing projects during development allows for the servicing of a customer base wider than the mining industry and the benefit of economies of scale. In the Lao People's

box continues next page

Box 5.6 Typology of Financing Arrangements *(continued)*

Democratic Republic, the Nam Theun II hydropower plant was designed with a capacity of 1,070 megawatts, 995 megawatts of which was dedicated for export to Thailand under a 25-year power purchase agreement with the Thai utility EGAT, at an agreed tariff on a take-or-pay basis. The revenue generated under this agreement with a creditworthy off-taker provided sufficient security to enable the financing of an expanded project, allowing an additional 75 megawatts of capacity to be dedicated for domestic consumption. Lenders to the project did not rely on the sales of electricity for the domestic market as part of their security package, but export sales did enable the project to bear modest oversizing and a limited additional capital cost. The project was commissioned in 2009 and is being operated for 25 years under a concession agreement, when it is to be transferred to the government of Lao PDR.

Financing infrastructure services indirectly related to the mining industry and supported by mining revenues. Project development with a ready and creditworthy off-taker as an anchor customer—such as a mine—can facilitate financing infrastructure serving the public. In Mozambique, the upstream development of natural gas fields in the Inhambane Province by Sasol, and the subsequent sale of the output under the Gas Sales Agreement to South Africa on a take-or-pay basis, allowed for the development of the gas pipeline connecting Mozambique with South Africa. ROMPCO, the company operating the pipeline, receives a transmission tariff from Sasol as the gas buyer and shipper through a ship-or-pay arrangement that is set at 80 percent of contract volume for 25 years. This guaranteed revenue could then serve to raise financing for the project, while third-party access is guaranteed by the regulator. Under the agreement stipulating pipeline use, open access is available for transportation and distribution of natural gas, within Mozambique from commissioning and within South Africa after 10 years. In Mozambique, the regulator approves third-party access provided there is uncommitted capacity after meeting the contracted amounts under the Gas Transportation Agreement. Ten years of restricted access to the South African pipeline allows Sasol the enhanced prospects for recovering costs and attracting the required financing.

Capitalizing mining revenue to fund infrastructure that is unrelated to the mining industry. Access to capital markets is limited in many sub-investment-grade countries, but allows financing of infrastructure beyond what is possible out of usual channels. Some revenue received from the mining industry via taxes, dividends, and/or royalties could potentially be converted from an annual income flow into a lump-sum capital stock through the use of a securitization approach. This could allow the funding of stand-alone transformational infrastructure supporting either the mining sector or another government priority sector, such as schools and hospitals. Borrowing against future receivables could allow projects to obtain an investment-grade credit rating—reducing financing cost and accessing a wider range of investors, provided the governance and management of capital funds that are created are structured well.

The following diagram (figure B5.6.1) outlines this concept more specifically. Part of the revenue flowing from the extractive industry to the government in royalties and taxes (1), or the respective state-owned enterprise via dividends, could be allocated to a designated escrow account or infrastructure fund (2). There, the revenue serves as collateral toward a loan issued by commercial lenders (3). The debt proceeds could then be allocated to the government budget for public works (option A); the state-owned enterprise responsible for

box continues next page

Box 5.6 Typology of Financing Arrangements *(continued)*

Figure B5.6.1 Capitalizing Mining Revenue to Fund Other Infrastructure

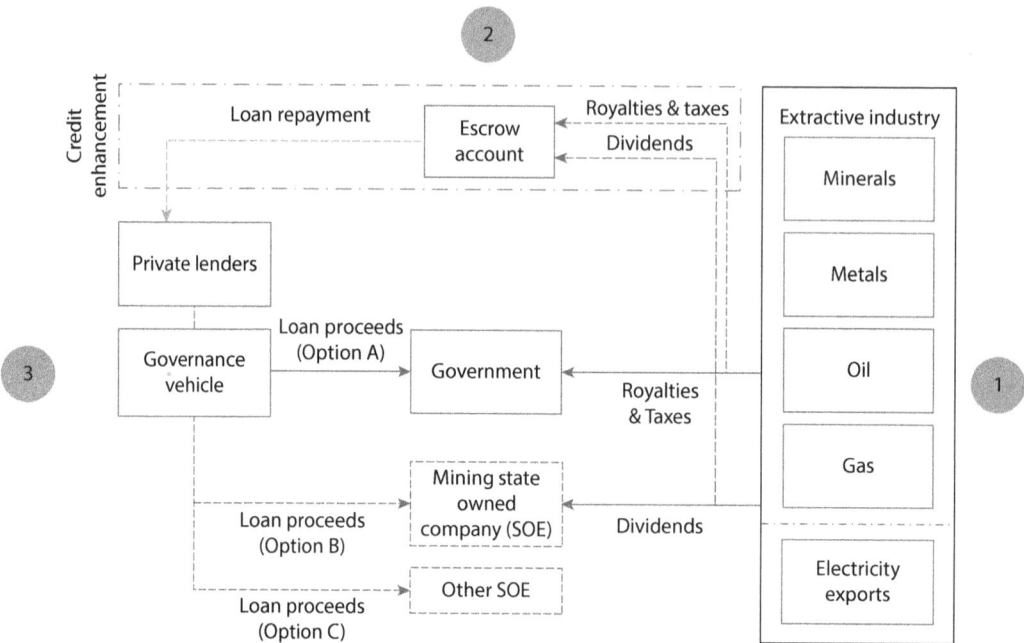

the extractives sector towards auxiliary infrastructure (option B); or, allocated to a separate government entity altogether (option C). An alternative would be to establish a separate fund management arrangement for the selection and preparation of projects before making investments, either alone or in partnership with other debt and equity providers.

Other benefits of securitization of extractive industry revenue include the mitigation of revenue uncertainty and volatility by passing on some of the risk of future revenue flows to private lenders and partially insulating the government from commodity price volatility, protecting revenues from politicization of investment management, and developing domestic capital market expertise and a track record benefitting future transactions.

Joint Development by Mines

For the smelter option in Mozambique (see chapter 4), it is envisaged that coal export companies will jointly develop the smelter and commit to supplying its power needs from discard coal generation over the smelter's 50-year life cycle. Over that period, replacement investment will be needed in the upstream power system. That the smelter owners also control the power supply meets one of the criteria for developing a smelter project. If a third party controlled the supply, renegotiating PPAs during the life of the smelter could be considered too great a risk for the owners. Thus, even if smelter power supplied by the large hydropower projects of Mphada Nkuwa and Cahora Bassa North has somewhat lower electricity costs, the smelter project may still opt for coal-fired supply.

An enduring private sector involvement in the power–mining domain in Zambia is Copperbelt Energy Corporation (CEC), a transmission and distribution company that buys bulk power from ZESCO and transmits and sells it to mines in the Copperbelt.

Notes

1. When that has been resolved, it may be possible to move ahead with the government's agenda of developing a wholesale electricity market to attract IPPs, with the mines as crucial, creditworthy off-takers to kick-start the process.

2. That is soon to be resolved; the key issue then is for the framework to be firmed up for the expected investments by mining companies in large hydropower projects.

3. But if the rail and port capacity is expanded as intended, power–mining integration will be facilitated, thus bringing additional benefits from the mining expansion.

4. These are the only countries among the eight that benefit from integrating the ministries.

References

Besant-Jones, John, Bernard Tenenbaum, and Prasad Tallapragada. 2008. *Regulatory Review of Power Purchase Agreements: A Proposed Benchmarking Methodology*. Washington, DC: World Bank Group. http://siteresources.worldbank.org/EXTAFRREGTOPENERGY /Resources/717305-1266613906108/ESMAP_337-08_Regulatory.pdf.

CEC (Copperbelt Energy Corporation). 2012. "Zambia Untapped Electricity Generation Potential." Lusaka, June 20, 2012.

Chanda, Gift. 2012. "CEC, AFC Sign $1.2bn Hydro-Power Project." *The Post Online*, March 16. http://www.postzambia.com/post-read_article.php?articleId=25881.

Creamer, Martin. 2011. "Questions stream in ahead of Anglo American Thermal Coal's IPP Deadline." *Mining Weekly*, October 19. http://www.miningweekly.com /article/reams-of-questions-stream-in-ahead-of-anglo-american-thermal-coals-ipp -deadline-2011-10-19.

Creamer, Terence. 2011. "First Power from Anglo American's Proposed Discard-Coal IPP TARGETED for 2015." *Mining Weekly*, July 14. http://www.miningweekly .com/article/first-power-from-anglo-americans-proposed-discard-coal-ipp-targeted -for-2015-2011-07-14-1.

Department of Energy. 2011. "Electricity Regulations on the Integrated Resource Plan 2010–2030." Republic of South Africa, Pretoria.

ERB (Energy Regulatory Board). 2011. "ERB Approves the ZESCO/CEC Application to Review Tariffs." Press Statement, August 9. Lusaka. http://www.erb.org.zm/press /statements/CECTariffApproval2011.pdf.

GBC News. 2008. "Mining Consortium to Transfer Generating Plant to VRA." June 9. http://www.modernghana.com/news/169010/1/mining-consortium-to-transfer -gnerating-plant-to-v.html.

Hall, Ian. 2010. "Khanyisa Project." Anglo American Integrated Resource Plan Public Hearing Presentation, Johannesburg, December 2. http://www.doe-irp.co.za/irpJHB /ANGLO_AMERICAN_THERMAL_COAL.pdf.

Harding, Claude. 2012. "South African Mines Might Start Generating Their Own Power." *How We Made It In Africa*, February 3. http://www.howwemadeitinafrica.com/south -african-mines-might-start-generating-their-own-power/14794/.

Hill, Matthew. 2011. "Semafo to Get Grid Power for Mana Mine." *Mining Weekly*, October 3. http://www.miningweekly.com/article/semafo-to-get-grid-power-for -mana-mine-2011-10-03.

Katanga Mining Limited. 2012. "Katanga Mining Signs Agreements to Develop Future Power Supply." Press Release, March 29. http://www.katangamining.com/media /news-releases/2012/2012-03-29.aspx.

Moyo, Roman. 2012. "Zimplats Lends ZESA US$25 Million." *New Zimbabwe*, August 4. http://www.newzimbabwe.com/business-8680-Zimplats+lends+ZESA+US$25+mill ion/business.aspx.

Nigerian Electricity Regulatory Commission. 2006. *Public Consultation on a Notice of Proposed Rulemaking (NOPR) for Review of PPAs to Supply Captive Customers.* Abuja.

Odendaal, Natasha. 2012. "Coal Miner Exxaro to Submit Five Renewable Energy Bids." *Engineering News*, February 23. http://www.engineeringnews.co.za/article/coal-miner -exxaro-to-submit-five-renewable-energy-bids-2012-02-23.

Othitis, Michael. 2012. "Independent NI 43-101 Technical Report: Turk and Angelus Mines, Zimbabwe, Africa." New Dawn Mining Corporation, Toronto. http://www.newdawnmining.com//wp-content/uploads/2013/04/2012-12-20 _TurkAngelusMines_NI_43-101.pdf.

Sibanda, Ntando. 2012. "Zesa Clears Cahora Bassa Debt, Gold Output Rises, Propping Up AirZim, Platinum Production." *Zimbabwe Development Democracy Trust News*, October 25. http://www.zddt.org/latest-zim-economic-news-1025.html.

Vale Columbia Center, Columbia University. 2012. "Leveraging the Mining Industry's Energy Demand to Improve Host Countries' Power Infrastructure." New York. http:// www.vcc.columbia.edu/files/vale/content/Working_paper-_Leveraging_the_mining _industrys_energy_demand_to_improve_host_countries_power_infrastructure _-_VCC.pdf.

World Bank. 2009. *Southern Mongolia Infrastructure Strategy.* Washington, DC. http:// siteresources.worldbank.org/INTMONGOLIA/Resources/SMIS_July.pdf.

———. 2011a. *Africa's Power Infrastructure: Investment, Integration, Efficiency.* Washington, DC.

———. 2011b. *Best Practices for Market-Based Power Rationing, Implications for South Africa.* Washington, DC.

———. 2012. "Cameroon—Lom Pangar Hydropower Project." Washington, DC. http:// siteresources.worldbank.org/INTCAMEROON/Resources/LPHP-PAD-Mar2012.pdf.

Zgodzinski, Dave. 2013. "How the Mining Sector Benefits from the Switch to Renewable Energy." *OilPrice.com*, March 19. http://oilprice.com/Finance/investing-and-trading -reports/How-the-Mining-Sector-Benefits-from-the-Switch-to-Renewable-Energy .html.

CHAPTER 6

A Way Forward

The future is all about potential: for mining companies to make steep power cost savings, for utilities to improve viability, and for the population to benefit from faster expansion of electricity access. The lack of power–mining integration that would bring widespread benefits to all parties is not due to mining companies' obduracy. On the contrary, the companies seem well aware of the potential for collaborative solutions with national utilities to help meet their objective of obtaining reliable, least-cost power.

The mining companies analyze their power supply options by country and project. As conditions change, they continually reassess their options and adapt. They offer the power sector considerable expertise in finding innovative solutions to a country's power problems, while also promoting their own interests. But the mines do not want to take on the role of a de facto utility—they establish clear boundaries around any assistance they provide.

The best solution for the mining companies would be strong national utilities that can efficiently deliver reliable, least-cost power, though they must negotiate tariffs based on the cost of supply so that the utility earns appropriate revenues. Otherwise, this approach impairs the utility's capacity to make new investments in extending connections, as well as broader power system development.

There are two other scenarios. The first is self-supply, in enclave or partial grid-connection variants, which entails heavy costs for the mining companies while undermining the national utility's viability by depriving it of major customers and making investments less viable because of lost scale economies. Mines may have the option to supply nearby communities that would not otherwise receive power. The second scenario focuses on relationships with mines selling collectively to the grid or mines serving as anchor demand for independent power producers (IPPs). If these involve merely leveraging the creditworthiness of the mining companies to finance power investments that would not otherwise materialize, they may not impose large costs. But they could be very expensive, in direct costs and opportunity costs of senior management time spent negotiating arrangements and overseeing project implementation. These two scenarios may nevertheless be worth pursuing as the basis for the power–mining projects explored in Guinea, Mauritania, and Tanzania.

Suggestions for Policymakers and the World Bank to Promote Power–Mining Integration

There is a need for greater strategic dialogue between the consolidated mining and power sectors in Sub-Saharan Africa (SSA) countries. To date, government interaction with mines appears to have been largely bilateral and focused on individual cases. As a result, there has been no opportunity for broader multistakeholder discussions between mining groups and government agencies responsible for power sector policy and planning. Such dialogue is needed to make progress in areas where there is good potential for solutions beneficial to all sides. The following recommendations are presented for policymakers, followed by potential areas of intervention by the World Bank.

Policymakers

Given the mining sector's power needs in many countries, governments could include current and projected mining power demand as part of their power sector master plans and demand-supply analysis and transmission planning once concrete agreements are in place. This would allow for a much clearer vision of a country's power situation and possibilities. To facilitate this integration, the ministry of mining (or relevant department where mining and energy are integrated in the same ministry, as in Cameroon, Mauritania, and Tanzania) should be consulted during the demand assessments.

Power requirements should be integrated into mining law. New mining operations should be required to outline their demand for power and sources. When collaboration is possible, the law should require dialogue between the mining company and relevant government agencies. The focus should be on dialogue, not mandated actions. For mines above a certain size that self-supply in remote areas, a requirement to supply local communities should be considered.

The national electricity sectors need to be sufficiently liberalized to at least allow for IPPs in the generation segment and, preferably, also encourage the private sector to invest in transmission. In all case study countries, the mines are big players: they have the capacity to invest directly in power generation and transmission or indirectly by serving as anchor customers for IPPs. The framework for IPP generation is already in place in all countries except the Democratic Republic of Congo, where new legislation is in the final stages of endorsement before being promulgated by the president. The future lies in incumbent utilities giving up their single-buyer monopsony to become nonexclusive central purchasers.

Carefully drafted contracts are important. Some mining companies insist on providing local electricity as part of their corporate social responsibility (CSR) policy, but this policy's voluntary nature can undermine its sustainability and national and local government responsibilities. Developing model concession agreements mandating electricity provision within a certain radius would increase certainty for investors, put all mining companies on an equal footing in their CSR programs, and increase the accountability of government as the contract enforcement authority.

Revenues generated by the mining sector via taxes, dividends, and/or royalties offer the potential to serve as an important driver of infrastructure development, if properly managed. Financial tools exist to facilitate delivery of public infrastructure services. Securitization can serve to improve borrowing conditions to investment grade and reduce financing cost; mitigate revenue uncertainty via risk sharing; promote strong governance, transparency, and solid asset management practices; and develop a capital market track record benefitting future transactions.

Governments should take a long-term perspective when identifying potential synergies and actions that will create an attractive enabling environment. Short-term solutions will often be incompatible with long-term country needs, particularly when rising demand from nonmining sources is taken into account. As is clear from the analysis here, many institutional arrangements are possible, and one-size-fits-all solutions should be avoided. Given the complexity of many situations, governments should seek expert advice to help identify the arrangements most likely to bring their countries the greatest benefits.

If a country's own short- and medium-term mining and nonmining power demand cannot justify what otherwise could be seen as an optimal power investment, a regional approach will be required to fully benefit from new arrangements. That will require regional coordination on policies and infrastructure related to power sharing.

Using large mining operations as anchor customers for large power development is attractive but should be undertaken with caution. It is important to have a flexible regime for power pricing to avoid subsidizing mines at the expense of the utility—or taxpayers. And the nonmining industries and residential consumers will eventually also want part of the power consumed by the mining company. Mining's large power demand will crowd out these other consumers, even if they are willing to pay a higher price. Contracts with the anchor customer must consider this possibility.

World Bank Group

Given the importance of power to mining and vice versa, the two sectors can no longer be viewed in isolation. World Bank staff should establish regular collaboration between the power and mining units, including joint missions in selected countries. They should also promote interaction between the country's power and mining stakeholders through national and local workshops and conferences to ensure that the various groups are aware of the possibilities, challenges, and responsibilities. Bank staff should also promote regional dialogue in harnessing large energy resources using mining companies as anchor consumers and supporting regional interconnection projects. In some subregions, this could be done through regional power pool platforms.

As a direct follow-up to this study, a toolkit could be developed that would include in-depth presentation and analysis of the various institutional arrangements that have been presented here, including methodologies for evaluating the costs and benefits of each. Finally, the wealth of information in the Africa Power–Mining Database will need to be regularly updated. The Bank should identify an African institution that could host this database.

Similar to the Extractive Industries Technical Advisory Facility, a trust fund should be considered to provide assistance to governments negotiating contracts between mining companies and power providers. Given the strong disadvantage of most African governments in such circumstances, this fund would level the playing field and arrive at arrangements in the countries' best interests.

The World Bank can directly engage in project ideas explored as scenarios in the analysis, particularly for countries with an inadequate and unreliable grid. These projects, which can be developed with mines as anchor consumers, will need government and donor support as mines have little incentive to create oversized generation plants to serve neighboring populations or send excess capacity to the grid.

The World Bank can also provide guidelines on how to better manage public–private partnerships (PPPs) and agreements on transmission, energy pricing, and retail distribution with the mines. A facility could be included to increase the capacity for effective application and adaptation to each country's institutional and regulatory situation. In certain cases, regulations may have already been enacted but are not being enforced. In addition, an initiative focusing on transparency of energy service agreements with the mines in both transmission and electricity pricing could perhaps be appended to already existing extractive industry transparency initiatives.

The Bank could provide support with convening power, technical inputs, and innovative financial instruments, such as guarantees and credit enhancements, to facilitate commercial arrangements among the mines and between the mines and utilities. Technical support can also assist in developing financing arrangements optimized toward local conditions. Where applicable the World Bank can facilitate securitization transactions to unlock the capital required for transformational projects and leverage its balance sheet to provide the necessary credit enhancements. Finally, in countries where the mines are largely grid supported, the Bank can conduct analytic studies and offer timely advice on rationalizing tariff structures and negotiating long-term mine tariffs.

Maps

Map A.1 Map of Energy Resources in Africa, by Location

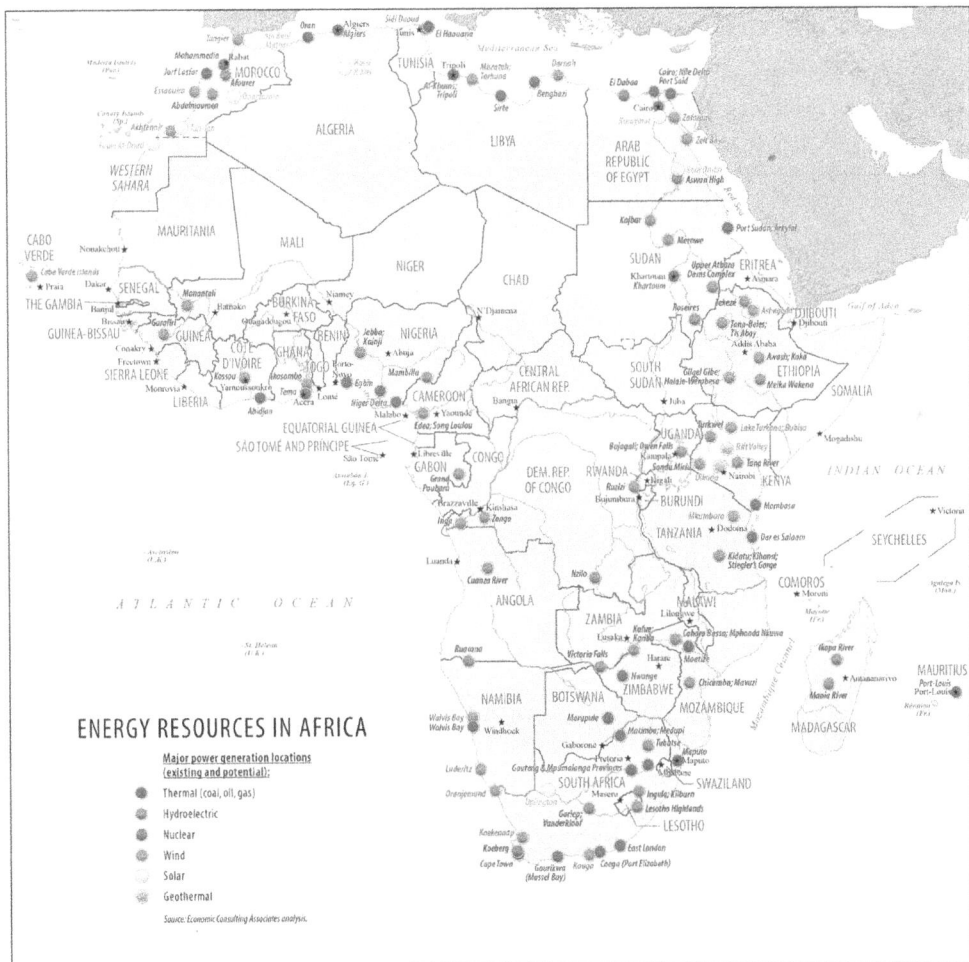

Source: World Bank, IBRD 40845, April 2014.

Map A.2 Localized Simulations of Power–Mining Synergies in Guinea

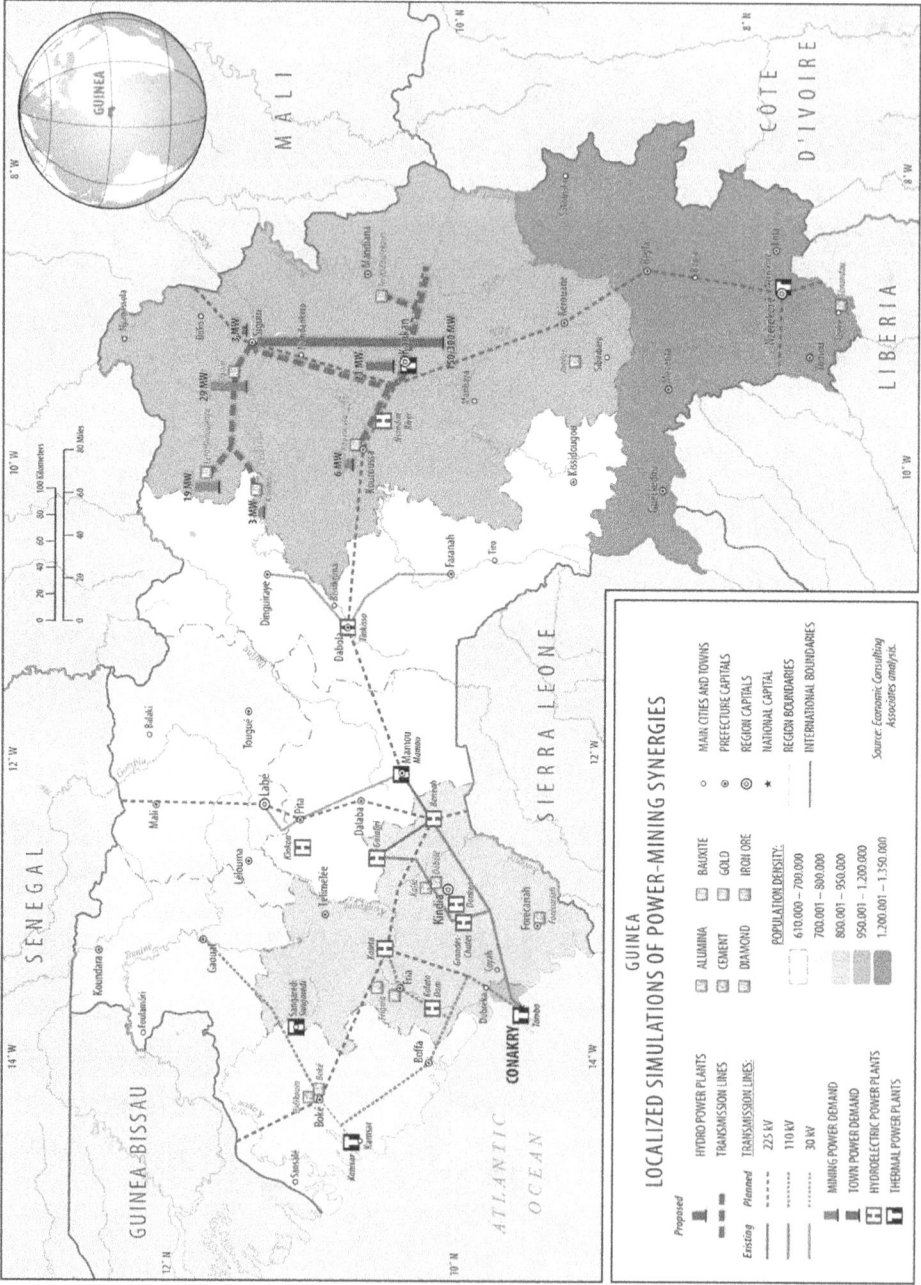

Source: World Bank, IBRD 40846, April 2014.

Map A.3 Localized Simulations of Power–Mining Synergies in Mauritania

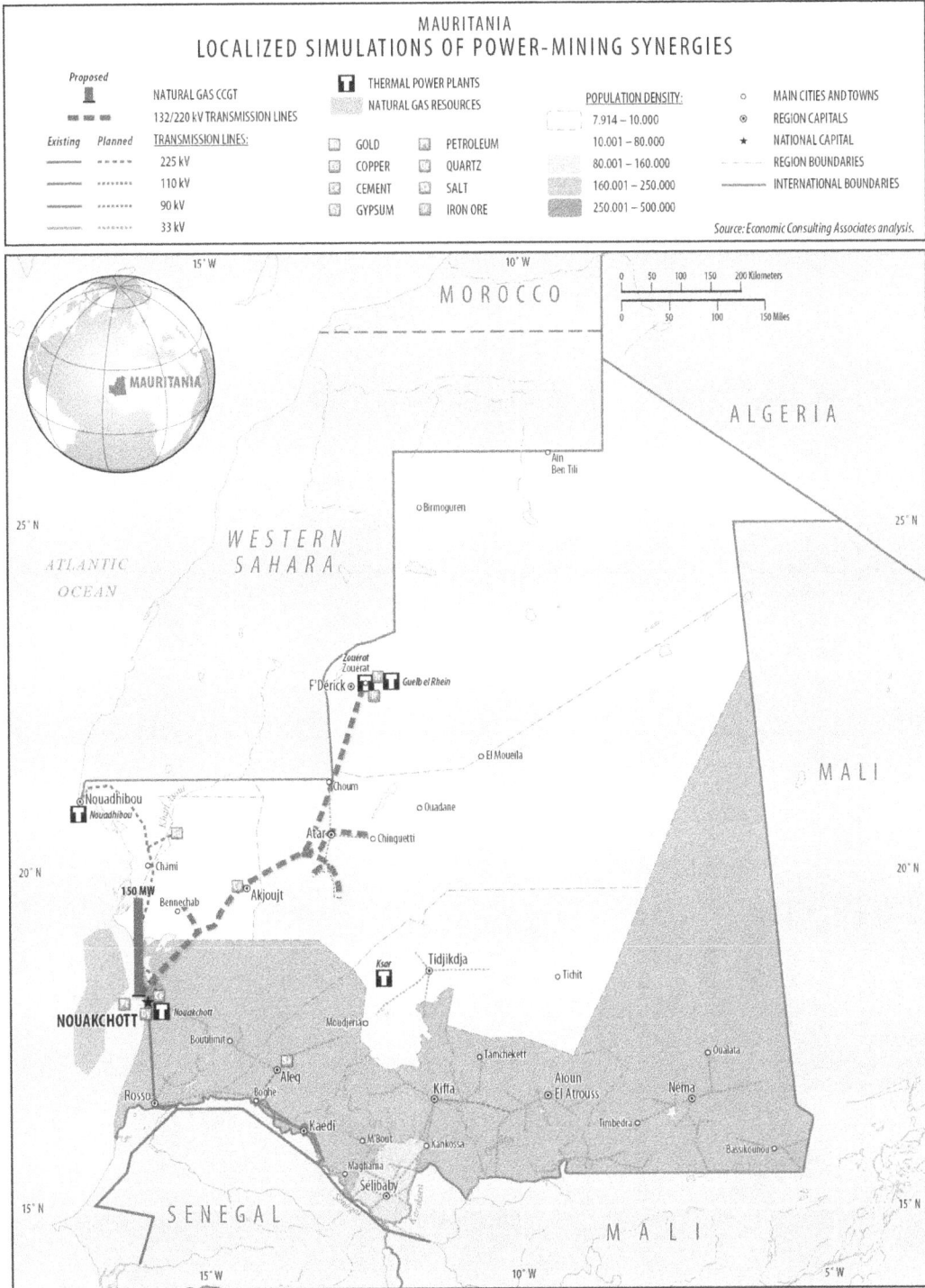

MAURITANIA
LOCALIZED SIMULATIONS OF POWER-MINING SYNERGIES

Proposed

NATURAL GAS CCGT
132/220 kV TRANSMISSION LINES

THERMAL POWER PLANTS
NATURAL GAS RESOURCES

POPULATION DENSITY:

7.914 – 10.000
10.001 – 80.000
80.001 – 160.000
160.001 – 250.000
250.001 – 500.000

MAIN CITIES AND TOWNS
REGION CAPITALS
NATIONAL CAPITAL
REGION BOUNDARIES
INTERNATIONAL BOUNDARIES

Existing Planned

TRANSMISSION LINES:
225 kV
110 kV
90 kV
33 kV

GOLD   PETROLEUM
COPPER   QUARTZ
CEMENT   SALT
GYPSUM   IRON ORE

Source: Economic Consulting Associates analysis.

Source: World Bank, IBRD 40847, March 2014.

The Power of the Mine • http://dx.doi.org/10.1596/978-1-4648-0292-8

Map A.4 Localized Simulations of Power–Mining Synergies in Tanzania

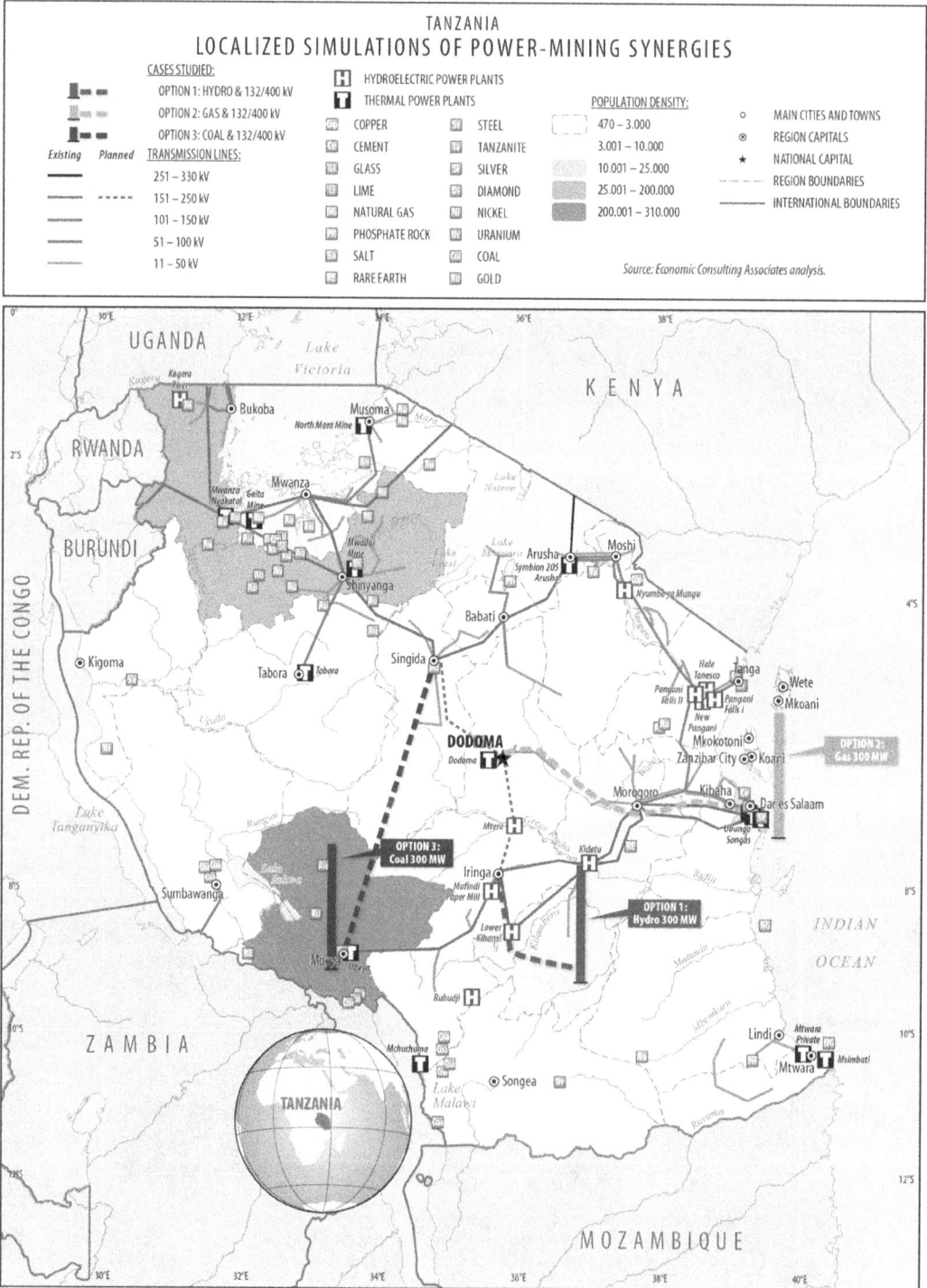

TANZANIA
LOCALIZED SIMULATIONS OF POWER-MINING SYNERGIES

CASES STUDIED:
OPTION 1: HYDRO & 132/400 kV
OPTION 2: GAS & 132/400 kV
OPTION 3: COAL & 132/400 kV

Existing   Planned

TRANSMISSION LINES:
251 – 330 kV
151 – 250 kV
101 – 150 kV
51 – 100 kV
11 – 50 kV

HYDROELECTRIC POWER PLANTS
THERMAL POWER PLANTS

COPPER   STEEL
CEMENT   TANZANITE
GLASS   SILVER
LIME   DIAMOND
NATURAL GAS   NICKEL
PHOSPHATE ROCK URANIUM
SALT   COAL
RARE EARTH   GOLD

POPULATION DENSITY:
470 – 3.000
3.001 – 10.000
10.001 – 25.000
25.001 – 200.000
200.001 – 310.000

MAIN CITIES AND TOWNS
REGION CAPITALS
NATIONAL CAPITAL
REGION BOUNDARIES
INTERNATIONAL BOUNDARIES

Source: Economic Consulting Associates analysis.

Source: World Bank, IBRD 40848, April 2014.

Map A.5 Localized Simulations of Power–Mining Synergies in Mozambique

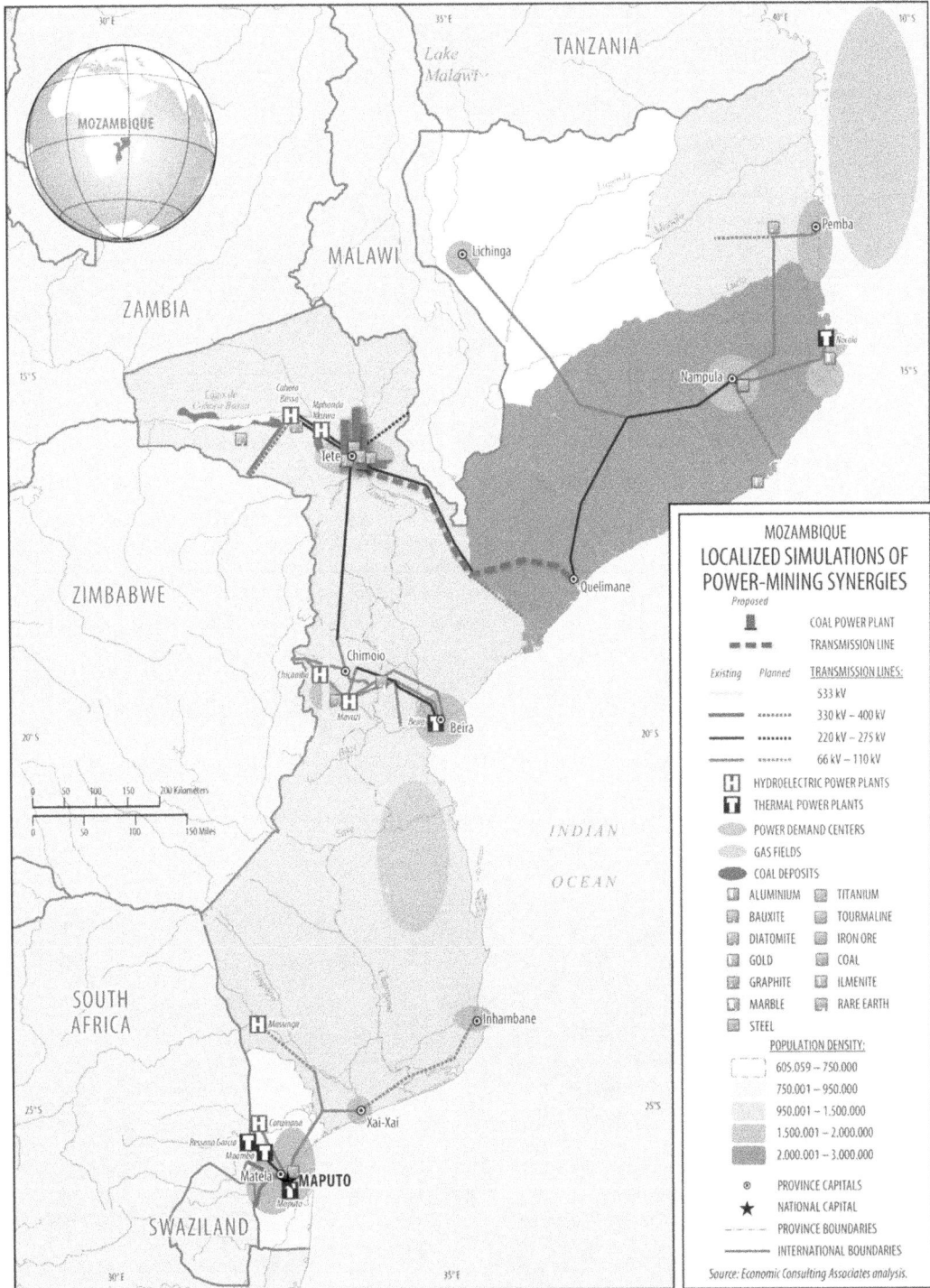

Source: World Bank, IBRD 40849, April 2014.

Map A.6 Cameroon Power–Mining Map

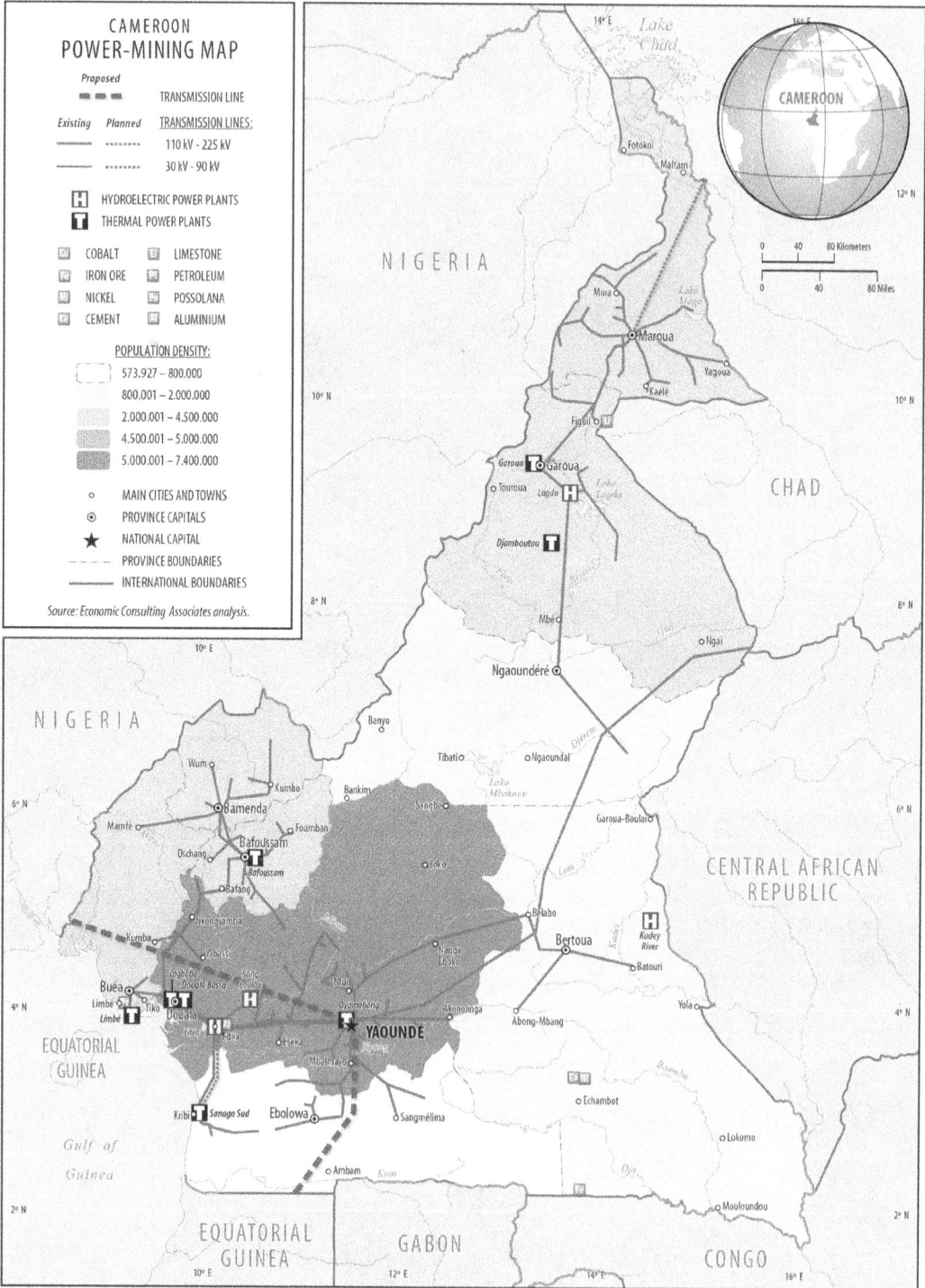

CAMEROON
POWER-MINING MAP

Proposed

TRANSMISSION LINE

Existing   *Planned*   TRANSMISSION LINES:

110 kV - 225 kV

30 kV - 90 kV

🄷 HYDROELECTRIC POWER PLANTS
🅃 THERMAL POWER PLANTS

COBALT   LIMESTONE
IRON ORE   PETROLEUM
NICKEL   POSSOLANA
CEMENT   ALUMINIUM

POPULATION DENSITY:

573.927 – 800.000
800.001 – 2.000.000
2.000.001 – 4.500.000
4.500.001 – 5.000.000
5.000.001 – 7.400.000

○   MAIN CITIES AND TOWNS
◉   PROVINCE CAPITALS
★   NATIONAL CAPITAL
–·–·– PROVINCE BOUNDARIES
—— INTERNATIONAL BOUNDARIES

Source: Economic Consulting Associates analysis.

CAMEROON

0   40   80 Kilometers
0   40   80 Miles

Lake Chad

14° E   16° E
12° N

NIGERIA

Fotokol
Maltam

Mora
Lake Maga
Maroua
Yagoua
Kaélé

Fiqui
Garoua
Geroua
Touroua
Laqdo
Lake Laqdo

CHAD

10° N   10° N

Djamboutou

8° N   8° N

Mbé
Ngai

Ngaoundéré

Banyo
Tibati   Ngaoundal
Lake
Mbakaou
Garoua-Boulaï

10° E

NIGERIA

Wum
Kumbo
Bankim
Bamenda
Foumban
Mamfé
Saangbe
Dschang   Bafoussam
Bafoussam
Bafang
Nkongsamba
Tibati
Kumba
Logbaba
Douala Bassa   Ntui
Buéa
Limbe Tiko
Limbé Douala
Edéa
Kribi Sanaga Sud

Belabo
Nanga
Eboko
Bertoua
Kadey
River
Batouri

CENTRAL AFRICAN
REPUBLIC

Yola

6° N   6° N

4° N   4° N

Oyomabong
YAOUNDÉ
Akonolinga
Abong-Mbang

EQUATORIAL
GUINEA

Mbalmayo
Pouma
Echambet
Lokomo

Ebolowa
Sangmélima

Gulf of
Guinea

Ambam

Mauloundou

EQUATORIAL
GUINEA   GABON   CONGO

2° N   2° N

10° E   12° E   14° E   16° E

Map A.7 Democratic Republic of Congo Power–Mining Map

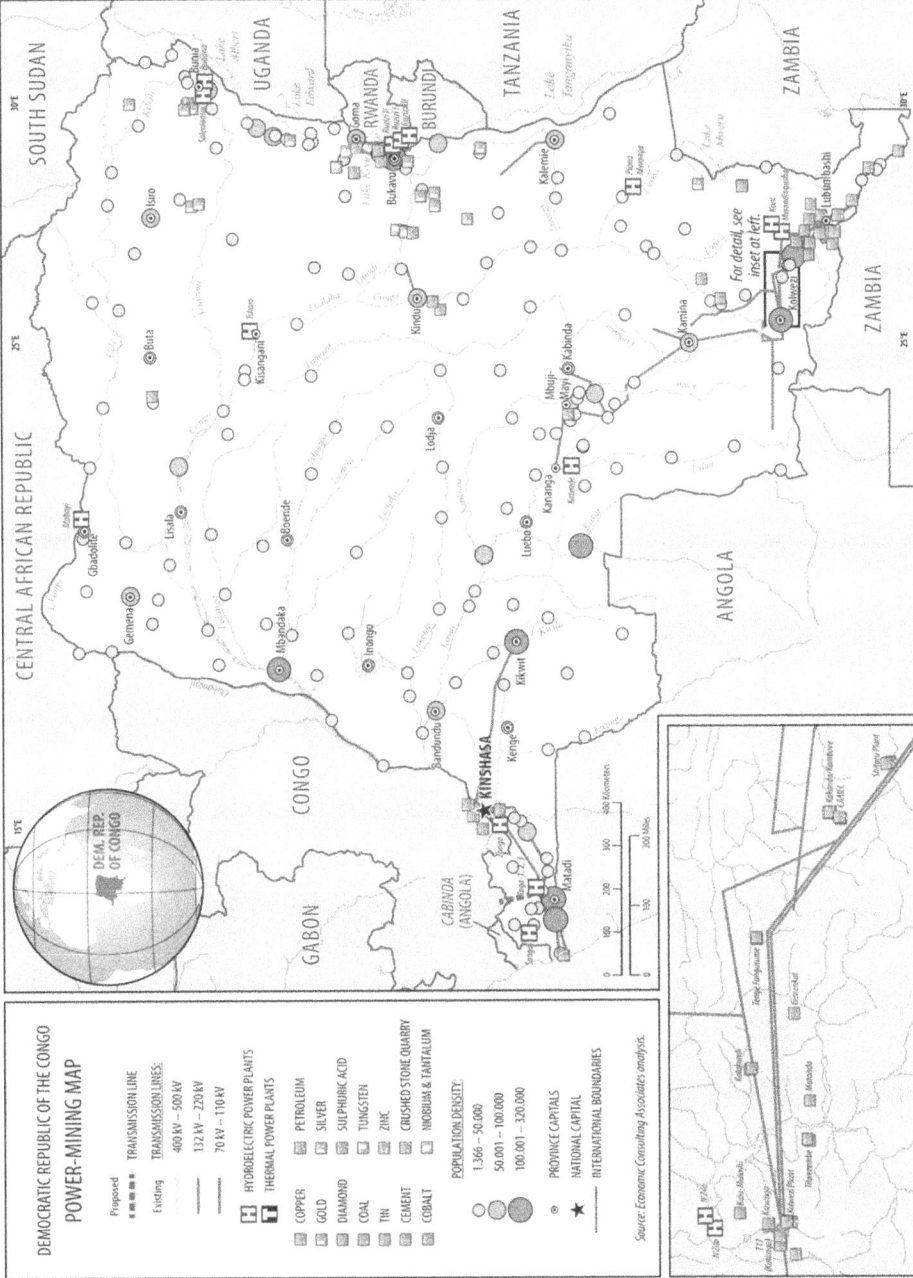

DEMOCRATIC REPUBLIC OF THE CONGO
POWER–MINING MAP

TRANSMISSION LINE
Proposed
Existing   TRANSMISSION LINES:
            400 kV – 500 kV
            132 kV – 220 kV
            70 kV – 110 kV

HYDROELECTRIC POWER PLANTS
THERMAL POWER PLANTS

COPPER   PETROLEUM
GOLD   SILVER
DIAMOND   SULPHURIC ACID
COAL   TUNGSTEN
TIN   ZINC
CEMENT   CRUSHED STONE QUARRY
COBALT   NIOBIUM & TANTALUM

POPULATION DENSITY:
1,365 – 50,000
50,001 – 100,000
100,001 – 320,000

PROVINCE CAPITALS
NATIONAL CAPITAL
INTERNATIONAL BOUNDARIES

Source: Economic Consulting Associates analysis.

Source: World Bank, IBRD 40851, April 2014.

109

Map A.8 Ghana Power–Mining Map

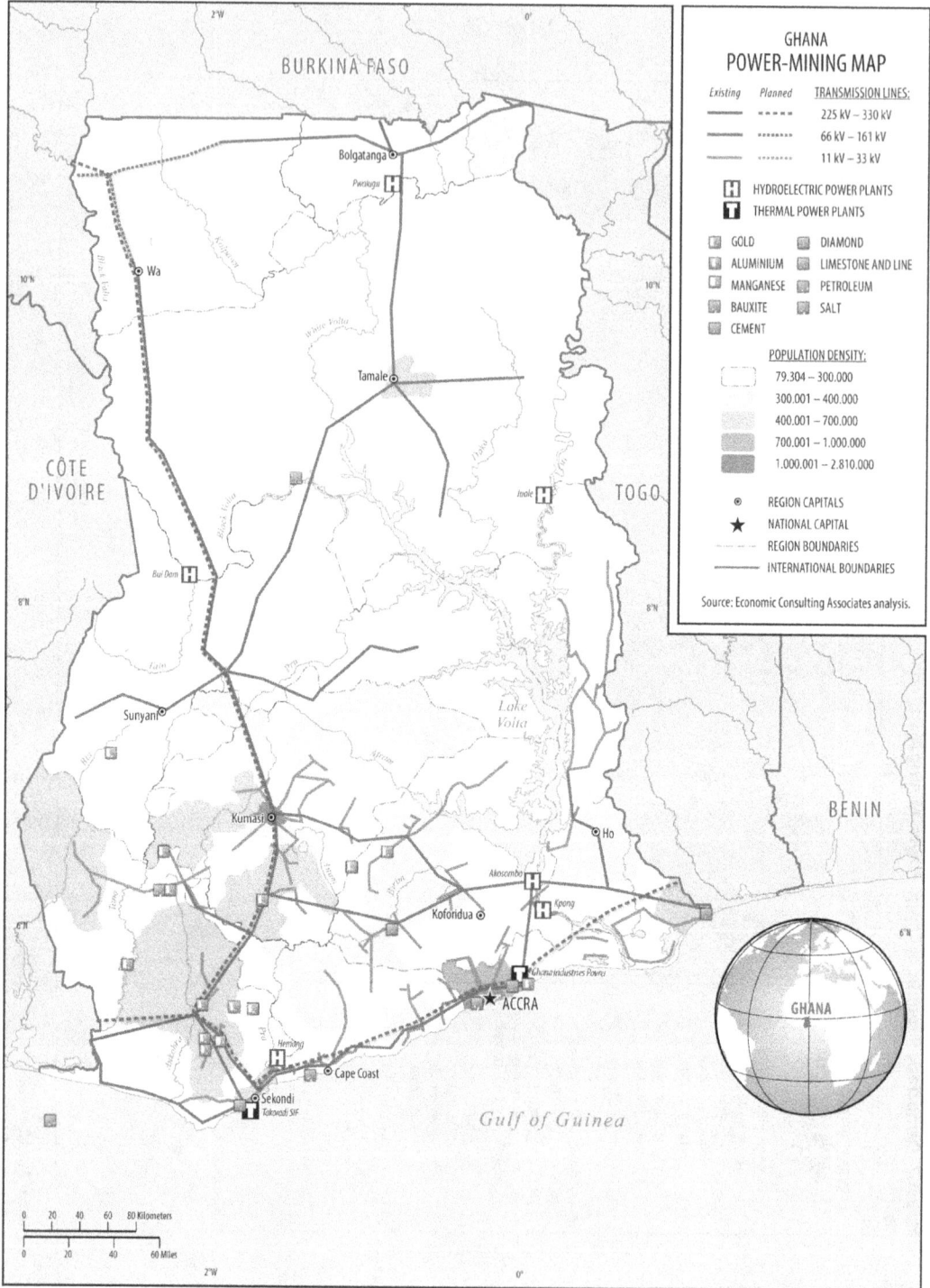

Source: World Bank, IBRD 40852, April 2014.

Map A.9 Zambia Power–Mining Map

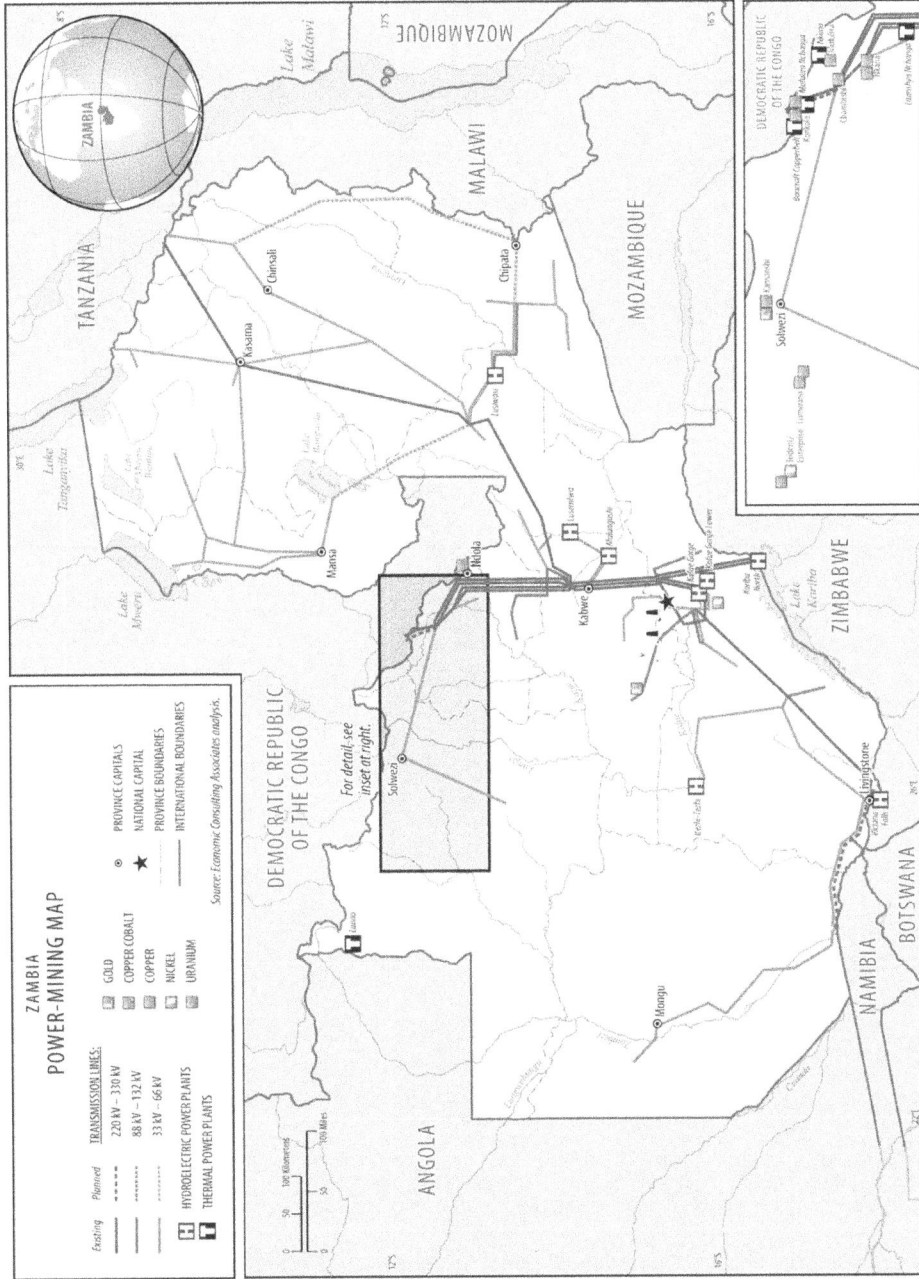

ZAMBIA
POWER–MINING MAP

Source: World Bank, IBRD 40853, April 2014.

111

Map A.10 Projects in the Africa Power–Mining Database

PROJECTS IN THE
AFRICA POWER-MINING DATABASE

Aluminum	Lead	Rare Earth
Bauxite	Manganese	Rhodium
Chromium	Nickel	Ruthenium
Coal	Nickel from N-C Sulphide	Rutile
Cobalt	Niobium	Silver
Copper	Palladium	Tantalum
Diamond	PGM	Uranium
Gold	Phosphate	Vanadium
Ilmenite	Platinum	Zinc
Iron Ore	Potash	Zirconium

ª Larger symbols indicate an aggregate of four mines located within 100 km of each other

Source: World Bank, IBRD 40917, April 2014.

Data Tables

Table B.1 Africa Power–Mining Database 2014

The database contains 455 projects and more than 60 fields. Below is a summary of the main fields explained in detail.

Main fields	Value			Comments
Type of work	Open-pit Open-pit/underground placer Plant	Surface Tailings Underground		Used to calculate the energy consumption for processing the ore
Status	Advanced exploration Closed Construction Development	Exploration Feasibility past producer Prefeasibility	Producer Raw project Temporary Suspension	
Country	28 SSA countries			SSA countries with mining operations
Commodity	Aluminum Bauxite Cement Chromium Coal Cobalt Copper Diamond Gold Ilmenite Iron ore	Lead Manganese Nickel Nickel from N-C sulphide Niobium Palladium PGM Phosphate Platinum Potash Rare earth	Rhodium Ruthenium Rutile Silico- manganese Silver Tantalum Uranium Vanadium Zinc Zirconium	
Best reserve available	Combined resources Indicated reserves Indicated resource Inferred resources Measured resource	Probable reserve Probable resources Proved and probable resources Proven and probable reserves Proven reserves		
Reserve grade				% grade of the ore
Current production				Production of the mine if already in operation

table continues next page

Table B.1 Africa Power–Mining Database 2014 (continued)

Main fields	Value		Comments
(Expected) yearly production at full capacity			Maximum production that the mine will be able to deliver. Used to calculate the annual energy consumption of the mine
Life of mine Project inception Project completion			Defines the lifespan of the mine and the time frame
Company name			Owner of the mine
Expected investment (US$ millions)			Expected investment needed to put the mine in operation
Exploitation phase Investment phase Feasibility phase Exploration phase	Complete Not started Temporary suspension Ongoing		Defines the time frame and the likelihood of moving to production by 2020
Probability to move to investment by 2020 (high, low)	High Low		Likelihood of moving to production by 2020
Extent of processing	Crushed Intermediate Processed	Smelted Refined Separation	Used to calculate the energy needs of the mines
Latitude, longitude			Coordinates of the mine
Region	Western Africa, Southern Africa, Eastern Africa, Central Africa		
Power-sourcing arrangement	A1, A2, B, C, D, E, F, G		Arrangement between the mine and the grid, ranging from self-supply to grid connection, with intermediate options
Source of power	Co-generation Co-generation/on-grid Grid Off-site coal Off-site gas On-grid, IPP On-grid, on-site HFO On-grid, on-site hydro (for equipment) On-grid, on-site solar On-grid, on-site solar On-site coal On-site coal, on-grid On-site diesel	On-site diesel, on-grid On-site diesel, off-site hydro On-site gas On-site gas, IPP On-site HFO On-site HFO, on-grid On-site HFO, on-site coal On-site HFO, on-site diesel On-site HFO, off-site gas On-site HFO, on-site hydro On-site hydro Off-site diesel, on-grid On-site diesel, on-site coal	Source of power used by the mine for its operations
Energy cost (c/kWh)	Min: 2 c/kWh for grid-connected mines in Lesotho Max: 29 c/kWh for on-site diesel generation Average: 11.07 c/kWh		The following assumptions were made: – Sourcing from off-site coal, gas, and hydro represents savings on own generation of 20% – For scenario where mines invest in grid, the energy cost used is the public utility tariff because the mine will invest in the infrastructure, which increases its costs but is compensated for through bill credits

table continues next page

Table B.1 Africa Power–Mining Database 2014 *(continued)*

Main fields	Value	Comments
Energy consumption (kWh/t product)	Min: 4 kWh/t for crushing bauxite Max: 69,489,780 kWh/t for refining palladium, PGM, platinum, rhodium, and ruthenium Average: 19,488,221 kWh/t	Calculated using the coefficients for each ore (see table B.4) and the field "extent of processing"
Annual energy consumption (MWh)	Min: 2 MWh Max: 14,200,000 MWh Average: 541,310 MWh	Calculated using "(expected) yearly production at full capacity" and "energy consumption kWh/t product"
Energy needs (MW)	Min: 0.5 MW (crushed manganese mine in Ivory Coast) Max: 1,479 MW (refined PGM in South Africa) Average: 56 MW	Energy needed by the mine to process the ore production at full capacity
Resource	CW, IM, WB, wise-uranium.org, Infomine, MW, USGS	Source of the information included in the database
Average price		Average price in US$/t for each ore
Threshold (exploration)	0/1	Most projects not currently in production have a low probability of moving into full-scale production by 2020 and have thus been eliminated from materiality, with exceptions for rare/precious mineral properties
Threshold nonzero	0/1	Properties are not considered material in instances where no reserve or production numbers are published by Infomine or USGS
Threshold (US$250 million)	0/1	Reserve assessed to be over US$250 million of initial investment
Projects–pre-2000 Projects–2001–12 Projects–2020 and on	Yes/no	Indicate whether the mine belongs to each of these time frames
Shareholding code	A, B, C, D, not awarded yet	
Tariff public utility	Min: 2 c/kWh in Lesotho Max: 53 c/kWh in Liberia Average: 9.1 c/kWh	Tariff for mining industry or industrial tariffs when available; otherwise average tariff Used to calculate the energy costs when the mine is connected to the grid
Cost relativity	Above tariff Below tariff Equal tariff	Compares the costs of energy generation incurred by the mine and the tariff in the country
Main source of generation	Mainly coal-fueled generation Mainly gas-powered generation Mainly hydro-powered generation Mainly oil-fueled generation	Main source of grid generation in the country
Grid reliability	0 to 50 days 100 to 150 days 50 to 100 days 150 to 200 days	Power outages (days per year, 2007–08)

Note: HFO = heavy fuel oil; IPP = independent power producer; PGM = platinum group metals; SSA = Sub-Saharan Africa.

Table B.2 Mining Demand—High, Low Probabilities

Country	Region	Number of projects	Pre-2000 (MW)	2012 (MW)	2020 high (MW)	2020 high and low (MW)
Angola	Southern Africa	9	4	4	101	101
Botswana	Southern Africa	20	126	142	134	134
Burkina Faso	Western Africa	15	9	62	133	133
Cameroon	Central Africa	6	145	145	224	1,112
Central African Republic	Central Africa	2	0	0	13	50
Congo, Rep.	Central Africa	4	0	0	200	200
Côte d'Ivoire	Western Africa	8	0.03	5	2	19
Congo, Dem. Rep.	Central Africa	37	35	587	917	1,149
Eritrea	Eastern Africa	7	0	65	98	98
Gabon	Central Africa	6	0	6	11	81
Ghana	Western Africa	12	67	273	369	369
Guinea	Western Africa	16	222	321	873	999
Kenya	Eastern Africa	3	0	0	88	88
Lesotho	Southern Africa	4	0	0.33	0.43	0.43
Liberia	Western Africa	8	0	31	86	222
Madagascar	Southern Africa	7	7	465	630	630
Malawi	Southern Africa	3	0	28	29	29
Mali	Western Africa	11	53	134	156	156
Mauritania	Western Africa	5	31	79	125	125
Mozambique	Southern Africa	8	807	1,120	1,310	1,310
Namibia	Southern Africa	15	88	156	480	480
Niger	Western Africa	6	98	103	241	241
Senegal	Western Africa	8	60	161	324	324
Sierra Leone	Western Africa	10	42	61	164	164
South Africa	Southern Africa	168	5,568	10,005	12,999	13,122
Tanzania	Eastern Africa	23	37	116	341	434
Zambia	Southern Africa	16	449	818	1,176	1,394
Zimbabwe	Southern Africa	18	145	237	266	279
Central Africa		**55**	**180**	**738**	**1,365**	**2,592**
Eastern Africa		**33**	**37**	**181**	**527**	**620**
Southern Africa		**268**	**7,195**	**12,975**	**17,124**	**17,479**
Western Africa		**99**	**582**	**1,230**	**2,473**	**2,752**
Total		**455**	**7,995**	**15,124**	**21,490**	**23,443**

Source: Africa Power–Mining Database 2014, World Bank, Washington, DC.

Table B.3 Mining Potential by Country (Energy Needs in MW)

Country/commodity	Number of projects	Pre-2000 (MW)	2001–12 (MW)	2020, high and low probability (MW)
Angola	**9**	**4**	**4**	**101**
Cement	1	0	0	26
Diamond	6	4	4	4
Iron ore	1	0	0	47
Phosphate	1	0	0	24
Botswana	**20**	**126**	**142**	**134**
Coal	2	18	18	18
Copper	4	38	53	70
Diamond	7	25	26	25
Gold	1	0	0	0
Nickel	1	41	41	0
Nickel from N-C sulphide	1	4	4	0
Platinum	1	0	0	2
Silver	2	0	0	0
Uranium	1	0	0	19
Burkina Faso	**15**	**9**	**62**	**133**
Gold	14	9	62	127
Zinc	1	0	0	6
Cameroon	**6**	**145**	**145**	**1,112**
Aluminum	2	145	145	484
Bauxite	1	0	0	549
Cobalt	1	0	0	4
Iron ore	1	0	0	73
Nickel	1	0	0	2
Central African Republic	**2**	**0**	**0**	**50**
Gold	1	0	0	13
Uranium	1	0	0	37
Congo, Rep.	**4**	**0**	**0**	**200**
Iron ore	3	0	0	156
Potash	1	0	0	43
Côte d'Ivoire	**8**	**0**	**5**	**19**
Gold	6	0	5	6
Manganese	2	0	1	14
Congo, Dem. Rep.	**37**	**35**	**587**	**1,149**
Cobalt	10	7	59	50
Copper	23	28	529	1,017
Gold	4	0	0	82
Eritrea	**7**	**0**	**65**	**98**
Copper	2	0	42	61
Gold	3	0	23	34

table continues next page

Table B.3 Mining Potential by Country (Energy Needs in MW) *(continued)*

Country/commodity	Number of projects	Pre-2000 (MW)	2001–12 (MW)	2020, high and low probability (MW)
Silver	1	0	0	0
Zinc	1	0	0	4
Gabon	**6**	**0**	**6**	**81**
Iron ore	1	0	0	42
Manganese	5	0	6	39
Ghana	**12**	**67**	**273**	**369**
Aluminum	1	65	65	65
Gold	10	0	206	302
Manganese	1	3	3	3
Guinea	**16**	**222**	**321**	**999**
Bauxite	6	174	271	272
Gold	4	48	51	55
Iron ore	6	0	0	672
Kenya	**3**	**0**	**0**	**88**
Ilmenite	1	0	0	69
Rutile	1	0	0	16
Zirconium	1	0	0	3
Lesotho	**4**	**0**	**0.3**	**0.4**
Diamond	4	0	0.3	0.4
Liberia	**8**	**0**	**31**	**222**
Gold	3	0	0	34
Iron ore	5	0	31	188
Madagascar	**7**	**7**	**465**	**630**
Chromium	2	7	7	5
Cobalt	1	0	0	4
Ilmenite	2	0	458	543
Nickel	1	0	0	35
Rutile	1	0	0	43
Malawi	**3**	**0**	**28**	**29**
Niobium	1	0	0	1
Tantalum	1	0	0	0
Uranium	1	0	28	28
Mali	**11**	**53**	**134**	**156**
Gold	11	53	134	156
Mauritania	**5**	**31**	**79**	**125**
Copper	1	0	18	18
Gold	1	0	30	30
Iron ore	3	31	31	77
Mozambique	**8**	**807**	**1,120**	**1,310**
Aluminum	1	807	807	807

table continues next page

The Power of the Mine · http://dx.doi.org/10.1596/978-1-4648-0292-8

Table B.3 Mining Potential by Country (Energy Needs in MW) *(continued)*

Country/commodity	Number of projects	Pre-2000 (MW)	2001–12 (MW)	2020, high and low probability (MW)
Coal	5	0	146	274
Ilmenite	1	0	167	167
Iron ore	1	0	0	63
Namibia	**15**	**88**	**156**	**480**
Copper	2	7	7	12
Diamond	1	0	0	1
Gold	2	0	0	18
Lead	1	2	2	2
Uranium	7	74	125	442
Zinc	2	6	23	6
Niger	**6**	**98**	**103**	**241**
Gold	2	0	5	4
Uranium	4	98	98	237
Senegal	**8**	**60**	**161**	**324**
Gold	3	0	11	42
Ilmenite	1	0	0	120
Phosphate	2	60	149	149
Rutile	1	0	0	6
Zirconium	1	0	0	7
Sierra Leone	**10**	**42**	**61**	**164**
Diamond	1	0	0	0
Gold	3	0	0	26
Ilmenite	1	0	0	4
Iron ore	3	0	19	92
Rutile	1	42	42	42
Zirconium	1	0	0	1
South Africa	**168**	**5,568**	**10,005**	**13,122**
Aluminum	2	1,330	1,330	1,330
Chromium	10	834	2,842	2,182
Coal	24	456	806	773
Copper	2	0	129	136
Diamond	7	8	9	8
Gold	38	785	1,075	904
Ilmenite	1	149	149	149
Iron ore	8	106	158	153
Lead	2	5	5	0
Manganese	8	28	212	638
Nickel from N-C sulphide	2	0	33	36
Palladium	4	137	154	187
PGM	10	385	735	2,341

table continues next page

The Power of the Mine • http://dx.doi.org/10.1596/978-1-4648-0292-8

Table B.3 **Mining Potential by Country (Energy Needs in MW)** *(continued)*

Country/commodity	Number of projects	Pre-2000 (MW)	2001–12 (MW)	2020, high and low probability (MW)
Platinum	33	443	1,280	1,692
Rare earth	2	0	0	25
Rhodium	2	31	36	36
Ruthenium	3	117	126	18
Rutile	1	11	11	11
Silico-manganese	2	0	143	481
Silver	1	0	0	0
Uranium	2	0	29	29
Vanadium	1	2	2	2
Zinc	1	1	1	0
Zirconium	2	739	739	1,990
Tanzania	**23**	**37**	**116**	**434**
Coal	2	0	2	13
Diamond	1	0	4	4
Gold	13	37	110	224
Iron ore	1	0	0	61
Nickel	4	0	0	66
Rare earth	1	0	0	30
Uranium	1	0	0	35
Zambia	**16**	**449**	**818**	**1,394**
Cobalt	2	2	2	2
Copper	10	447	795	1,357
Gold	2	0	15	26
Nickel from N-C sulphide	2	0	6	8
Zimbabwe	**18**	**145**	**237**	**279**
Chromium	1	81	81	81
Coal	3	17	17	77
Diamond	2	0	13	13
Gold	7	10	70	41
Nickel	2	0	1	13
PGM	2	0	54	54
Platinum	1	36	0	0
Total	**455**	**7,995**	**15,124**	**23,443**

Source: Africa Power–Mining Database 2014, World Bank, Washington, DC.
Note: Totals may not sum due to rounding. PGM = platinum group metals.

Table B.4 Coefficients (kWh/mt of Product)

Commodity	Crushed	Crushed UG	Intermediate	Intermediate UG	Smelted	Smelted UG	Refined	Refined UG	Processed	Processed UG	Separation	Separation UG
Aluminum	—	—	—	—	15,500	—	—	—	—	—	—	—
Bauxite	4	—	—	—	—	—	620	—	—	—	—	—
Cement	38	—	45	—	—	—	—	—	125	—	—	—
Chromium	100	200	340	440	4,340	4,640	—	—	—	—	—	—
Coal	37	55	—	—	1,337	1,355	—	—	—	—	—	—
Cobalt	—	—	6,500	7,000	7,550	8,050	8,150	8,650	—	—	—	—
Copper	—	—	5,000	7,000	6,050	8,050	6,650	8,650	—	—	—	—
Diamond	—	—	—	—	—	—	—	—	—	—	30,000,000	37,500,000
Gold	—	—	—	—	—	—	25,000,000	45,000,000	—	—	—	—
Ilmenite	—	—	2,000	2,200	4,100	4,300	9,300	9,500	—	—	—	—
Iron ore	20	30	60	70	590	600	—	—	—	—	—	—
Lead	—	—	855	855	1,015	1,015	1,183	1,183	—	—	—	—
Manganese	16	30	—	—	2,516	2,530	—	—	—	—	—	—
Nickel	2,000	3,000	5,600	6,600	—	—	9,700	10,700	—	—	—	—
Nickel from N-C sulphide	—	—	300	550	1,100	1,350	3,100	3,350	—	—	—	—
Niobium	—	—	—	—	—	—	4,000	—	—	—	—	—
Palladium	—	—	28,219,200	47,619,900	46,561,680	65,962,380	50,089,080	69,489,780	—	—	—	—
Phosphate	75	—	—	—	—	—	—	—	287	287	—	—

table continues next page

Table B.4 Coefficients (kWh/mt of Product) *(continued)*

Commodity	Crushed	Crushed UG	Intermediate	Intermediate UG	Smelted	Smelted UG	Refined	Refined UG	Processed	Processed UG	Separation	Separation UG
Platinum	—	—	28,219,200	47,619,900	46,561,680	65,962,380	50,089,080	69,489,780	—	—	—	—
PGM	—	—	28,219,200	47,619,900	46,561,680	65,962,380	50,089,080	69,489,780	—	—	—	—
Potash	60	60	460	460	—	—	630	630	—	—	—	—
Rare earth	—	—	—	—	—	—	—	—	—	—	7,200	8,000
Rhodium	—	—	28,219,200	47,619,900	46,561,680	65,962,380	50,089,080	69,489,780	—	—	—	—
Ruthenium	—	—	28,219,200	47,619,900	46,561,680	65,962,380	50,089,080	69,489,780	—	—	—	—
Rutile	—	—	2,000	2,200	4,100	4,300	9,300	9,500	—	—	—	—
Silico-manganese	16	30	—	—	4,016	4,030	—	—	—	—	—	—
Silver	—	—	—	—	—	—	9,645	9,645	—	—	—	—
Steel	—	—	—	—	—	—	790	790	—	—	—	—
Tantalum	—	—	—	—	—	—	4,000	—	—	—	—	—
Uranium	—	—	48,000	48,000	—	—	178,000	178,000	—	—	—	—
Vanadium	—	—	—	—	—	—	3,000	3,000	—	—	—	—
Zinc	—	—	595	595	1,050	1,050	1,075	1,075	—	—	—	—
Zirconium	—	—	800	900	54,600	54,700	59,800	59,900	—	—	—	—

Source: USGS 2011.

Note: PGM = platinum group metals; UG = underground; — = not available.

Table B.5 Annual Energy Consumption by Country (MWh)

Country	Pre-2000	2001–12	2020 (high and low probability)
Angola	38,970	40,386	966,896
Botswana	1,214,263	1,364,663	1,286,336
Burkina Faso	89,301	593,635	1,273,910
Cameroon	1,395,000	1,395,000	10,677,570
Central African Republic	0	0	483,573
Congo, Rep.	0	0	1,915,800
Côte d'Ivoire	325	52,222	186,872
Congo, Dem. Rep.	336,400	5,637,342	11,288,978
Eritrea	0	621,835	944,138
Gabon	0	54,400	775,520
Ghana	644,000	2,620,385	3,541,743
Guinea	2,129,169	3,083,975	9,592,134
Kenya	0	0	842,000
Lesotho	0	3,174	4,134
Liberia	0	300,000	2,130,272
Madagascar	62,900	4,462,900	6,044,900
Malawi	0	266,440	280,440
Mali	508,292	1,287,903	1,499,523
Mauritania	300,000	758,495	1,198,495
Mozambique	7,750,000	10,756,000	12,577,000
Namibia	848,625	1,494,312	4,605,102
Niger	943,400	987,342	2,313,837
Senegal	574,000	1,541,311	3,107,505
Sierra Leone	400,000	583,240	1,575,510
South Africa	53,453,747	96,047,459	125,968,895
Tanzania	354,369	1,114,964	4,166,138
Zambia	4,314,750	7,854,076	13,383,013
Zimbabwe	1,392,330	2,273,084	2,681,805
Total	**76,749,839**	**145,194,541**	**225,312,037**

Source: Africa Power–Mining Database 2014, World Bank, Washington, DC.

Table B.6 Average Annual Energy Consumption by Country (MWh)

Country	Pre-2000	2001–2012	2020 (high and low probability)
Angola	12,990	6,731	161,149
Botswana	134,918	104,974	91,881
Burkina Faso	89,301	148,409	115,810
Cameroon	1,395,000	1,395,000	1,779,595
Central African Republic	0	0	241,786
Congo, Rep.	0	0	478,950
Côte d'Ivoire	325	13,055	31,145
Congo, Dem. Rep.	84,100	256,243	418,110
Eritrea	0	310,917	134,877
Gabon	0	54,400	129,253
Ghana	322,000	291,154	295,145
Guinea	354,861	385,497	639,476
Kenya	0	0	280,667
Lesotho	0	1,058	1,034
Liberia	0	300,000	266,284
Madagascar	31,450	1,487,633	1,007,483
Malawi	0	266,440	93,480
Mali	169,431	183,986	166,614
Mauritania	300,000	252,832	239,699
Mozambique	7,750,000	2,689,000	1,572,125
Namibia	212,156	249,052	328,936
Niger	471,700	329,114	462,767
Senegal	574,000	513,770	388,438
Sierra Leone	400,000	194,413	157,551
South Africa	722,348	774,576	893,396
Tanzania	354,369	159,281	181,136
Zambia	862,950	872,675	892,201
Zimbabwe	278,466	189,424	191,557
Average total	**604,329**	**560,597**	**574,776**

Source: Africa Power–Mining Database 2014, World Bank, Washington, DC.

Table B.7 Detailed Mining Projects for the Eight Case Study Countries

Commodity	Mine	Type of mine	Estimated yearly production (t)	Project inception	Project completion	Probability to move to investment by 2020	Energy needs (MW)	Energy cost (c/kWh)	Power-sourcing arrangement	Source of power
a. Cameroon										
Aluminum	Alucam	Plant	210,000	2017	—	Low	339	4	B	On-site hydro
Aluminum	Alucam	Plant	90,000	1960	2050	High	145	16	G	On-grid
Bauxite	Ngaoundal/Minitrap	Surface	8,500,000	2019	—	Low	549	4	B	On-site hydro
Cobalt	Mada/Nkamouna	Open-pit	6,100	2014	2038	High	4	29	A1	On-site diesel
Iron ore	Mbalam	Surface	35,000,000	2016	2041	High	73	14.5	A1, F	On-site diesel, off-site hydro
Nickel	Mada/Nkamouna	Open-pit	3,200	2014	2038	High	2	29	A1	On-site diesel

table continues next page

125

Table B.7 Detailed Mining Projects for the Eight Case Study Countries (continued)

Commodity	Mine	Type of mine	Estimated yearly production (t)	Project inception	Project completion	Probability to move to investment by 2020	Energy needs (MW)	Energy cost (c/kWh)	Power-sourcing arrangement	Source of power
b. Congo, Dem. Rep.										
Cobalt	Kakanda/Kambove Tailings	Tailings	3,500	1982	1997	Closed	2	4.7	E	On-grid
Cobalt	Kakanda North/South	Surface	3,500	—	—	Closed	2	4.7	G	On-grid
Cobalt	Kalukundi	Open-pit/underground	3,800	2013	2020	High	3	4.7	G	On-grid
Cobalt	Kamoto (Katanga)	Underground	9,900	2007	2030	High	7	4.7	E	On-grid
Cobalt	Kananga (Dcp)	Open-pit	5,688	2004	2006	Closed	4	4.7	E	On-grid
Cobalt	Kov (Katanga)	Open-pit	17,000	2008	2030	High	12	4.7	E	On-grid
Cobalt	Mutanda	Open-pit	23,000	2010	2030	High	16	4.7	E	On-grid
Cobalt	T17 (Katanga)	Open-pit/underground	6,200	2007	2025	High	5	4.7	E	On-grid
Cobalt	Tenke Fungurume	Open-pit	13,000	2009	2024	High	9	4.7	E	On-grid
Cobalt	Tilwezembe Mine	Open-pit	7,100	1999	2008	Closed	5	4.7	E	On-grid
Copper	Deziwa/Ecaille Copper	Surface	200,000	2015	—	Low	126	4.7	G	On-grid
Copper	Frontier (Lufua)	Open-pit	70,000	2007	2029	High	36	4.7	G	On-grid
Copper	Kakanda/Kambove Tailings	Tailings	45,000	1982	1997	Closed	23	4.7	E	On-grid
Copper	Kakanda North/South	Surface	53,000	—	—	Low	28	4.7	G	On-grid
Copper	Kalukundi	Open-pit/underground	16,400	2013	2020	High	12	4.7	G	On-grid
Copper	Kalumines	Surface	20,000	2008	2009	Closed	10	4.7	E	On-grid
Copper	Kamoa	Surface	150,000	2017	2078	Low	78	4.7	E	On-grid
Copper	Kamoto (Katanga)	Underground	75,000	2007	2030	High	55	4.7	E	On-grid
Copper	Kananga (Dcp)	Open-pit	11,600	2004	2006	Closed	6	4.7	E	On-grid
Copper	Kansuki	Surface	90,000	—	—	High	47	4.7	E	On-grid

table continues next page

Table B.7 Detailed Mining Projects for the Eight Case Study Countries *(continued)*

Commodity	Mine	Type of mine	Estimated yearly production (t)	Project inception	Project completion	Probability to move to investment by 2020	Energy needs (MW)	Energy cost (c/kWh)	Power-sourcing arrangement	Source of power
b. Congo, Dem. Rep.										
Copper	Kinsenda	Open-pit/underground	26,000	2015	2035	High	19	4.7	G	On-grid
Copper	Kinsevere	Open-pit	60,000	2007	2023	High	31	4.7	E	On-grid
Copper	Kipoi	Open-pit	39,000	2011	2023	High	20	9.6	C	On-site diesel, on-grid
Copper	Kolwezi	Tailings	70,000	2013	2033	High	36	4.7	G	On-grid
Copper	Kolwezi Concentrator	Open-pit	30,954	2008	2061	High	16	4.7	G	On-grid
Copper	Kov (Katanga)	Open-pit	200,000	2008	2030	High	104	4.7	E	On-grid
Copper	Kulu–Mutoshi	Open-pit	57,000	2006	2008	Closed	30	4.7	G	On-grid
Copper	Lonshi	Open-pit	30,000	2001	2007	Closed	16	0	0	—
Copper	Mutanda	Open-pit	110,000	2010	2030	High	57	4.7	E	On-grid
Copper	Sicomine	Surface	400,000	2015	2045	High	208	4.7	E	On-grid
Copper	T17 (Katanga)	Open-pit/underground	52,000	2007	2025	High	38	4.7	E	On-grid
Copper	Tenke Fungurume	Open-pit	200,000	2009	2024	High	104	4.7	E	On-grid
Copper	Tilwezembe Mine	Open-pit	8,500	1999	2008	Closed	4	4.7	E	On-grid
Gold	Kibali (Moto Gold Project)	Open-pit/underground	17	2013	2029	High	41	4	A2	On-site hydro
Gold	Mobale/Kamituga	Open-pit/ underground	0.3	2015	2023	High	11	29	A1	On-site diesel
Gold	Mongbwalu	Surface	4.3	2014	2033	High	11	29	A1	On-site diesel
Gold	Namoya	Open-pit/underground	3.5	2013	2020	High	19	29	A1	On-site diesel

table continues next page

Table B.7 Detailed Mining Projects for the Eight Case Study Countries *(continued)*

c. Ghana

Commodity	Mine	Type of mine	Estimated yearly production (t)	Project inception	Project completion	Probability to move to investment by 2020	Energy needs (MW)	Energy cost (c/kWh)	Power-sourcing arrangement	Source of power
Aluminum	Tema	Plant	40,000	1970	—	High	65	11.1	G	On-grid
Gold	Ahafo/Subika–Ntotoroso	Open-pit/underground	17.6	2006	2021	High	83	17.15	D, G	Off-site diesel, on-grid
Gold	Akoase	Surface	2.8	2015	2025	High	7	11.1	G	On-grid
Gold	Akyem	Surface	13.6	2013	2028	High	35	17.15	D, G	Off-site diesel, on-grid
Gold	Bibiani Mine	Open-pit	1.5	2011	2024	High	4	11.1	G	On-grid
Gold	Chirano	Open-pit/underground	3.5	2005	—	High	16	11.1	G	On-grid
Gold	Damang/Abosso	Open-pit/underground	6.2	2011	2025	High	29	17.15	D, G	Off-site diesel, on-grid
Gold	Idupapriem Gold Mine	Open-pit	5.7	2006	2018	High	15	17.15	D, G	Off-site diesel, on-grid
Gold	Konongo	Surface	0.3	2009	2023	High	1	11.1	G	On-grid
Gold	Obuasi Mine (Ashanti)	Open-pit/underground	11.3	2013	2033	High	53	17.15	D, G	Off-site diesel, on-grid
Gold	Tarkwa Mine	Open-pit	22.5	2011	2024	High	58	17.15	D, G	Off-site diesel, on-grid
Manganese	Nsuta	Open-pit	1,500,000	1916	2027	High	3	11.1	G	On-grid

table continues next page

Table B.7 Detailed Mining Projects for the Eight Case Study Countries *(continued)*

Commodity	Mine	Type of mine	Estimated yearly production (t)	Project inception	Project completion	Probability to move to investment by 2020	Energy needs (MW)	Energy cost (c/kWh)	Power-sourcing arrangement	Source of power
d. Guinea										
Bauxite	Bidikoum	Surface	14,000,000	1992	—	High	6	29	A2	On-site diesel
Bauxite	Boke Bauxite Mine	Open-pit	12,068,000	1968	2043	High	5	29	A2	On-site diesel
Bauxite	Debele (Kindia Mining Complex)	Open-pit	3,200,000	1972	—	High	1	13.55	A1, G	On-site diesel, on-grid
Bauxite	Friguia	Open-pit	2,500,000	1960	2023	High	161	29	A2	On-site diesel
Bauxite	Koba–Koumbia Bauxite (Amig)	Surface	3,200,000	2016	—	Low	1	20	A1	On-site HFO
Bauxite	Sangaredi	Open-pit	1,500,000	2007	2038	High	97	29	A1	On-site diesel
Gold	Kiniero Mine	Open-pit	1	2002	2015	Closed by 2020	3	14.5	A1, D	On-site diesel, off-site hydro
Gold	Lefa (Dinguiraye)	Open-pit	7.4	1995	—	High	19	29	A2	On-site diesel
Gold	Siguiri Mine	Surface	11.1	1998	2022	High	29	29	A2	On-site diesel
Gold	Tri-K/Koulékoun	Underground	0	2015	2022	High	7	29	A2	On-site diesel

table continues next page

129

Table B.7 Detailed Mining Projects for the Eight Case Study Countries (continued)

Commodity	Mine	Type of mine	Estimated yearly production (t)	Project inception	Project completion	Probability to move to investment by 2020	Energy needs (MW)	Energy cost (c/kWh)	Power-sourcing arrangement	Source of power
d. Guinea										
Iron ore	Euronimba	Surface	30,000,000	2018	2038	High	63	18.25	A1	On-site diesel, on-site coal
Iron ore	Forecariah	Underground	5,000,000	2013	2021	High	16	29	A1	On-site diesel
Iron ore	Kalia	Surface	50,000,000	2013	—	High	313	20	A1	On-site HFO
Iron ore	Mount Nimba	Surface	10,000,000	2017	2027	Low	21	29	A1	On-site diesel
Iron ore	Simandou (Rio)	Surface	75,000,000	2018	2068	High	156	13.55	A2, E	On-Site diesel, on-grid
Iron ore	Simandou (Vale)—block 1 and 2	Surface	50,000,000	2019	2040	Low	104	29	A1	On-site diesel

table continues next page

Table B.7 Detailed Mining Projects for the Eight Case Study Countries *(continued)*

Commodity	Mine	Type of mine	Estimated yearly production (t)	Project inception	Project completion	Probability to move to investment by 2020	Energy needs (MW)	Energy cost (c/kWh)	Power-sourcing arrangement	Source of power
e. Mauritania										
Copper	Guelb Moghrein Mine	Open-pit	35,000	2006	2021	High	18	29	A1	On-site diesel
Gold	Tasiast	Open-pit	11.3	2008	2046	High	30	12.596	A1, D	On-site HFO, off-site gas
Iron ore	Askaf	Surface	6,000,000	2014	—	High	13	29	A1	On-site diesel
Iron ore	Guelb El Aouj	Surface	16,000,000	2017	2040	High	33	25	E	On-grid
Iron ore	Zouerate	Open-pit	15,000,000	1975	2040	High	31	12.596	A1, D	On-site HFO, off-site gas

table continues next page

131

Table B.7 Detailed Mining Projects for the Eight Case Study Countries (continued)

f. Mozambique

Commodity	Mine	Type of mine	Estimated yearly production (t)	Project inception	Project completion	Probability to move to investment by 2020	Energy needs (MW)	Energy cost (c/kWh)	Power-sourcing arrangement	Source of power
Aluminum	Mozal	Plant	500,000	2000	2050	High	807	4.1	G	On-grid
Coal	Benga/Tete	Surface	16,000,000	2012	2037	High	62	7.5	B	On-site coal
Coal	Chirodzi	Open-pit	10,000,000	2013	2038	High	39	7.5	B	On-site coal
Coal	Estima	Open-pit	20,000,000	2016	2041	High	77	7.5	B	On-site coal
Coal	Minas Moatize	Open-pit	3,000,000	2014	2028	High	12	7.5	B	On-site coal
Coal	Moatize Coal	Open-pit	22,000,000	2011	2046	High	85	7.5	B	On-site coal
Ilmenite	Moma	Surface	800,000	2004	2132	High	167	4.1	E	On-grid
Iron ore	Tete/Ruoni	Underground	1,000,000	2016	2046	High	63	4.1	G	On-grid

table continues next page

132

Table B.7 Detailed Mining Projects for the Eight Case Study Countries *(continued)*

Commodity	Mine	Type of mine	Estimated yearly production (t)	Project inception	Project completion	Probability to move to investment by 2020	Energy needs (MW)	Energy cost (c/kWh)	Power-sourcing arrangement	Source of power
g. Tanzania										
Coal	(Mbalawala Mine) Ngaka Thermal Coal	Open-pit	500,000	2011	2026	High	2	7.5	B	On-site coal
Coal	Mchuchuma	Open-pit	3,000,000	2015	—	Low	12	7.5	B	On-site coal
Diamond	Williamson (Mwadui)	Open-pit	1.3	2010	—	High	4.2	5.3	G	On-grid
Gold	Buckreef MI	Underground	1.8	2014	2023	High	8	7.65	C	On-site HFO, on-grid
Gold	Bulyanhulu	Underground	7.4	2009	2034	High	35	12.65	E, A1	On-grid, on-site HFO
Gold	Buzwagi	Open-pit	5.6	2009	2022	High	15	5.3	E	On-grid
Gold	Geita Gold Mine	Open-pit	14.2	2000	2022	High	37	7.65	A1, E	On-site HFO, on-grid
Gold	Golden Ridge Project	Open-pit	2.8	2014	2020	High	7	7.65	C	On-site HFO, on-grid
Gold	Lupa Goldfield/Luika	Surface	2	2012	2032	High	5	7.65	C	On-site HFO, on-grid
Gold	Magambazi	Surface	2.3	2016	2026	Low	6	20	A1	On-site HFO

table continues next page

133

Table B.7 Detailed Mining Projects for the Eight Case Study Countries *(continued)*

Commodity	Mine	Type of mine	Estimated yearly production (t)	Project inception	Project completion	Probability to move to investment by 2020	Energy needs (MW)	Energy cost (c/kWh)	Power-sourcing arrangement	Source of power
g. Tanzania										
Gold	Miyabi	Surface	2	2016	2026	High	5	7.65	C	On-site HFO, on-grid
Gold	Mwanza	Underground	5.7	2016	—	High	27	20	A1	On-site HFO
Gold	North Mara	Open-pit	7.1	2002	2021	High	18	5.3	E	On-grid
Gold	Nyanzaga	Underground	11.3	2016	2026	High	53	7.65	C	On-site HFO, on-grid
Gold	Saza Makongolosi Project	Open-pit	0.4	—	—	High	1	7.65	C	On-site HFO, on-grid
Gold	Singida	Underground	1.3	—	—	High	6	7.65	C	On-site HFO, on-grid
Iron ore	Liganga	Open-pit	1,000,000	2015	—	Low	61	7.5	A1	On-site coal
Nickel	Dutwa	Underground	27,000	2016	2036	High	19	7.65	C	On-site HFO, on-grid
Nickel	Kabanga	Surface	40,000	2013	2043	High	23	20	A1	On-site HFO

table continues next page

Table B.7 Detailed Mining Projects for the Eight Case Study Countries *(continued)*

Commodity	Mine	Type of mine	Estimated yearly production (t)	Project inception	Project completion	Probability to move to investment by 2020	Energy needs (MW)	Energy cost (c/kWh)	Power-sourcing arrangement	Source of power
g. Tanzania										
Nickel	Nachingwea/Ntaka Hill	Underground	15,000	2016	—	High	10	7.65	C	On-site HFO, on-grid
Nickel	Ngwena	Underground	20,000	2016	2035	Low	14	9.9	A1, G	On-site diesel, on-grid
Rare earth	Ngualla	Surface	40,000	2016	2041	High	30	20	A1	On-site HFO
Uranium	Mkuju River/Nyota	Underground	1.905	2013	2025	High	35	20	A1	On-site HFO

table continues next page

135

Table B.7 Detailed Mining Projects for the Eight Case Study Countries *(continued)*

h. Zambia

Commodity	Mine	Type of mine	Estimated yearly production (t)	Project inception	Project completion	Probability to move to investment by 2020	Energy needs (MW)	Energy cost (c/kWh)	Power-sourcing arrangement	Source of power
Cobalt	Nkana (Mopani)	Open-pit/underground	3,000	1937	2036	High	2	3.9	G	On-grid
Cobalt	Sable (Glencore)	Plant	26	2009	2019	High	0	3.9	G	On-grid
Copper	Baluba	Underground	40,000	2010	—	High	36	3.9	G	On-grid
Copper	Chambishi	Underground	200,000	1965	1987	Closed	180	3.9	G	On-grid
Copper	Kansanshi	Open-pit/underground	400,000	2005	2021	High	360	3.9	G	On-grid
Copper	Konkola	Underground	200,000	2005	—	High	168	3.9	G	On-grid
Copper	Konkola Deep	Underground	200,000	2014	—	High	168	3.9	G	On-grid
Copper	Konkola North	Underground	100,000	2015	2031	High	90	3.9	G	On-grid
Copper	Lumwana	Open-pit	150,000	2015	2052	High	95	3.9	G	On-grid
Copper	Mufulira (Mopani)	Underground	135,000	1937	2043	High	122	3.9	G	On-grid
Copper	Nkana (Mopani)	Open-pit/underground	150,000	1937	2036	High	109	3.9	G	On-grid
Copper	Sentinel/Kalumbila	Underground	250,000	2015	2035	Low	210	3.9	G	On-grid
Gold	Kansanshi	Open-pit/underground	3.2	2005	2021	High	15	3.9	G	On-grid
Gold	Luiri Hill Project	Open-pit/underground	2.4	—	—	High	11	3.9	G	On-grid

table continues next page

Table B.7 Detailed Mining Projects for the Eight Case Study Countries *(continued)*

Commodity	Mine	Type of mine	Estimated yearly production (t)	Project inception	Project completion	Probability to move to investment by 2020	Energy needs (MW)	Energy cost (c/kWh)	Power-sourcing arrangement	Source of power
h. Zambia										
Nickel from N-C sulphide	Munali (Enterprise/ Voyager Deposits)	Underground	100,000	2008	2018	Low	6	3.9	G	On-grid
Nickel from N-C sulphide	Trident/ Enterprise	Surface	85,000	2015	—	Low	3	3.9	G	On-grid

Source: Africa Power–Mining Database 2014, World Bank, Washington, DC.

Note: Power-sharing arrangements:

A1: Self-supply.

A2: Self-supply + CSR.

B: Self-supply + sell to the grid.

C: Grid supply + self-supply backup.

D. Mines sell collectively to the grid.

E: Mines invest in the grid.

F: Mines serve as anchor demand for IPP.

G: Grid supply.

Cells with A2, B, C, D, E, or F indicate intermediate arrangements; cells with double-letter entries (for example, "A1, F" or "D, G") indicate transitional arrangements; — = not available.

Table B.8 Socioeconomic Details for Eight Case Study Countries

Country	Population (millions)	GDP per capita (PPP, US$)	GDP growth rate (%)	Life expectancy at birth (years)	Mortality rate under 5 (per 1,000)	Adult literacy rate (%)	Electrification rate (%)	Electricity consumption per capita (kWh p.a.)
	2011	2011	2011	2011	2011	2010	2009	2010
Guinea	10.2	1,124	3.9	64	78	67	18	79
Congo, Dem. Rep.	67.8	373	6.9	52	127	77	11	95
Tanzania	46.2	1,512	6.4	58	68	73	14	78
Mozambique	23.9	975	7.1	50	103	58	12	444
Mauritania	3.5	2,554	4.8	49	83	71	19	444
Ghana	25.0	1,871	14.4	54	126	41	61	298
Cameroon	20.0	2,359	4.2	50	103	56	49	271
Zambia	13.5	1,621	6.5	48	168	67	19	623

Source: World Development Indicators 2011 (database), World Bank, Washington, DC, http://data.worldbank.org /data-catalog/world-development-indicators/wdi-2011.
Note: PPP = purchasing power parity.

Table B.9 Institutional Details for Eight Case Study Countries

		Electricity market						Financial and management condition of public utility				
	Market structure	Regulator	Generation	Transmission/ distribution	Electricity sector related authorities	Main regulatory framework		Very poor	Poor	Balanced	Well	Very well
Country						Name	Details					
Guinea	State-owned vertically integrated company with private owned power plants allowed to sell electricity to grid	Ministry of Energy and Hydraulics/ CNEE	EDG CBG ACG	EDG	DNE CNEE	Law L/92/043 (1992) Law L/92/043 (1993)	Related to the Code of Economic Activities Regulates the responsibility frameworks of the various stakeholders, the management and operation modalities of power generation, transmission and distribution activities, the provisions for participation by the private sector allocation of concessions, creation of the CNEE, and the penalty provisions applicable to the power sector	✓				
						Law L/97/012 (1998)	Regulates the financing, construction, operation, maintenance, and transfer of production facilities developed by private operators: BOT (Built, Operate, and Transfer), BOO (Built, Own, and Operate), BOOT (Build, Own, Operate, Transfer)					

Source: Guinea Article IV March 2012_ IMF 2012, Guinea letter of intent Sept 2012_IMF 2012

table continues next page

Table B.9 Institutional Details for Eight Case Study Countries (continued)

Country	Market structure	Regulator	Generation	Transmission/ distribution	Electricity sector related authorities	Main regulatory framework — Name	Main regulatory framework — Details	Very poor	Poor	Balanced	Well	Very well
Congo, Dem. Rep.	Vertically integrated utility	Ministry of Energy	SNEL	SNEL		Electricity code	Policy in place to encourage private players (to liberalize the electricity sector—creating free and fair codes of competition, protecting both users and operators)				✓	
						Rural electrification strategy	Targeting increased private sector involvement in rural electrification					
							Source: Congo, Dem. Rep. Infrastructures_2010					
Tanzania	Vertically integrated utility	EWURA	TANESCO Songas Artumas Energy IPLT TANWAT TPC Wentworth Power Limited	TANESCO		Electricity Act 2008	Primary legislation applicable to the generation and/or transmission and/or distribution of electrical power in Tanzania and to provide for cross-border trade in electricity and rural electrification		✓			
						EWURA Act	Establishes the Energy and Water Utilities Regulator Authority (EWURA)					
							Source: TANESCO Financial Statement_2011					

table continues next page

Table B.9 Institutional Details for Eight Case Study Countries *(continued)*

Country	Market structure	Regulator	Generation	Transmission/distribution	Electricity sector related authorities	Main regulatory framework Name	Main regulatory framework Details	Financial and management condition of public utility Very poor	Poor	Balanced	Well	Very well
Mozambique	Vertically integrated on the way to unbundling	Ministry of Energy	HCB EDM	EDM HCB MoTraCo	CNELEC	DNEE	Entitles the Ministry of Energy as the supervisor of renewable energy				✓	
						DNER	Entitles the Ministry of Energy as the regulator and the supervisor of electricity sector					
						Law No. 21/97 (1997)	Electricity law					
						Decree No. 8/2000 (2000)	Regulations on the National Power Transmission Network Regulations					
						Decree No. 43/2005 (2005)	Entrusts the role of NPTN Operator to Electricidade de Moçambique					
						Decree No. 45/98 (1998)	Regulations on management of power facilities built or renovated with own funds in the districts that have not been assigned to a public company					

table continues next page

141

Table B.9 Institutional Details for Eight Case Study Countries *(continued)*

Country	Electricity market							Financial and management condition of public utility				
	Market structure	Regulator	Generation	Transmission/ distribution	Electricity sector related authorities	Main regulatory framework		Very poor	Poor	Balanced	Well	Very well
						Name	Details					
						Ministerial Dipl. No. 31/85 (1985)	Regulations on Technical Skills for preparing, implementing, and operating power facilities of particular service					
						Decree No. 48/2007 (2007)	Licensing regulations for electric facilities					
						Decree No. 25/2000 (2000)	Electricity National Council (CNELEC) Statutes					
							Source: Mozambique Infrastructure_AICD 2011					
Mauritania	Vertically integrated monopoly	Authority of regulation	SOMELEC SOGEM	SOMELEC		2001 Electricity code	Promotes the liberalization of the sector → effective only in rural areas. Self-sufficient energy producers have an estimated capacity of 40 MW		✓			
Ghana	Liberalized market	PURC EC	VRA TICO Takoradi II Sunon Asogli Bui hydro	GridCo ECD NED							✓	
							Source: Ghana Infrastructure_AICD 2010					

table continues next page

Table B.9 Institutional Details for Eight Case Study Countries *(continued)*

Country	Market structure	Electricity market				Main regulatory framework		Financial and management condition of public utility				
		Regulator	Generation	Transmission/ distribution	Electricity sector related authorities	Name	Details	Very poor	Poor	Balanced	Well	Very well
Cameroon	Vertically integrated monopoly	ARSEL	AES Sonel	AES Sonel	EDC	New Electricity Law (2011)	Encouraging private investments in hydropower and creation of a public transmission company to unbundle the sector		✓			
							Source: Cameroon Infrastructure_AICD 2010					
Zambia	Effective monopoly	ERB	ZESCO CEC Lunsemfwa Hydro Power Company	ZESCO CEC					✓			
							Source: Zambia Infrastructure_AICD 2010.					

Source: Dominguez-Torres and Briceno-Garmendia 2011; Dominguez-Torres and Foster 2011; Foster and Benitez 2011; Foster and Dominguez 2011; Foster and Pushak 2011.

Table B.10 Power Sector Statistics for Eight Case Study Countries

Country/year		Installed capacity (MW)	Operational capacity (MW)	Operational/ installed (%)	Electricity net generation (GWh)	Electricity net consumption (GWh)	Peak demand (MW)	Installed capacity factor	Operational capacity factor	Net exports (GWh)	Electricity transmission losses (GWh)	Electricity distribution losses (GWh)
Guinea	(2010)	395	88	22	969	901	120	0.28	1.26	0		68
Congo, Dem. Rep.	(2010)	2,437	1,170	48	7,800	6,197	1,081	0.37	0.76	755	753	850
Tanzania	(2010)	1,115	880	79	4,300	3,400	833	0.44	0.56	−50		900
Mozambique	(2010)	2,428	2,250	93	16,499	10,212	560	0.78	0.84	4,787	4,784	1,503
Mauritania	(2010)	263	—	—	701	652	—	0.30	—	0		49
Ghana	(2011)	2,169	1,909	88	11,200	7,976	1,547	0.59	0.67	610	531	1,920
Cameroon	(2010)	1,115	980	88	5,761	5,181	—	0.59	0.67	0		580
Zambia	(2010)	1,679	1,215	72	11,192	7,960	1,600	0.76	1.05	545	545	2,687

Source: ECA Consulting Group data.

Note: Electricity transmission and distribution losses are aggregated for Guinea, Tanzania, Mauritania, and Cameroon; — = not available.

References

Dominguez-Torres, Carolina, and Cecilia Briceno-Garmendia. 2011. "Mozambique's Infrastructure: A Continental Perspective." Policy Research Working Paper 5885, World Bank, Washington, DC.

Dominguez-Torres, Carolina, and Vivien Foster. 2011. "Cameroon's Infrastructure: A Continental Perspective." Policy Research Working Paper 5822, World Bank, Washington, DC.

Foster, Vivien, and Daniel Alberto Benitez. 2011. "The Democratic Republic of Congo's Infrastructure: A Continental Perspective." Policy Research Working Paper 5602, World Bank, Washington, DC.

Foster, Vivien, and Carolina Dominguez. 2011. "Zambia's Infrastructure: A Continental Perspective." Policy Research Working Paper 5599, World Bank, Washington, DC.

Foster, Vivien, and Nataliya Pushak. 2011. "Ghana's Infrastructure: A Continental Perspective." Policy Research Working Paper 5600, World Bank, Washington, DC.

USGS (United States Geological Survey). 2011. "Estimates of Electricity Requirements for the Recovery of Mineral Commodities, with Examples Applied to Sub-Saharan Africa." U.S. Department of Energy, Washington, DC.

Methodology

This appendix presents the methodology for simulation of localized scenarios computed for Guinea, Mauritania, Mozambique, and Tanzania.

Mining power demand. The primary sources on mining power demand are from the Africa Power–Mining Database 2014 developed under this study. This database identifies demand for power from mining since 2000 and forecasts demand for power from projects that may become operational by 2020 in 28 Sub-Saharan Africa (SSA) countries. This work was carried out by the Columbia Center on Sustainable Investment, a joint center of Columbia Law School and the Earth Institute at Columbia University. It was reviewed and revised by the World Bank team and SRK Consulting for the case study countries.

Dispatch model. The estimated short- and long-run marginal costs (LRMCs) are based on a model of economic dispatch developed as part of a separate project by Economic Consulting Associates (ECA) for the World Bank evaluation of gas-to-power demand in the 47 SSA countries. A brief description is presented here, while a full description of this model is found in Santley, Schlotterer, and Eberhard (2013). In modeling economic dispatch, country-level supply and demand and opportunities for trade are evaluated.

On the demand side, a daily load curve is developed for each country based on information provided in the regional and national master studies, with the underlying energy demand based on forecasts prepared in each of the power pool master plans. The choice of the power pool demand forecasts follows careful evaluation of the various information sources (national power sector development plans, regional plans, and the Program for Infrastructure Development in Africa study). To create the daily load curve, three load blocks—base, mid, and peak—are generated to capture the role of the load shape on economic dispatch of the supply portfolio.

Although it was originally intended to capture seasonal variation, evaluation of temperature and load data revealed that most SSA countries did not experience seasonal variation in demand, though this is not to say that all countries have no seasonal variation—a case in point is South Africa. However, for the handful of countries with seasonality, there was not enough information to appropriately inform the modeling. So no seasonal variation has been modeled. For each of the three load blocks, average demand for a given hour in the block is to be identified and used to determine the corresponding generation dispatch in the same hour.

On the supply side, a plant-level supply curve is constructed, based on all available generation in a given year (as defined by the least-cost investment sequence from the power development plan being applied). The cost of production (based, for gas thermal, on the prices developed by the upstream consultant) for each plant, along its corresponding energy production, is then ordered from the lowest to highest cost.

To capture the role of hydro seasonality, plus the ability to shape the water to meet peak load, three supply curves (peak, high, and low) are constructed for each country with hydro in its generation portfolio.

For each country-level generation portfolio and country-level demand, five dispatch states are modeled:

- *Peak*—hydro capacity is rated at the peak hydro capacity and the dispatch optimized to meet peak demand in a least-cost manner.
- *High hydro/mid demand*—hydro capacity is rated at the high hydro capacity and the dispatch optimized to meet the mid demand block in a least-cost manner.
- *High hydro/base demand*—hydro capacity is rated at the high hydro capacity and the dispatch optimized to meet the base demand block in a least-cost manner.
- *Low hydro/mid demand*—hydro capacity is rated at the low hydro capacity and the dispatch optimized to meet the mid demand block in a least-cost manner.
- *Low hydro/base demand*—hydro capacity is rated at the low hydro capacity and the dispatch optimized to meet the base demand block in a least-cost manner.

Geo-referenced linkages. To analyze the alternative scenarios, it was necessary to include the investments for transmission infrastructure to interconnect the mines and to identify the potential demand from neighboring residential consumers in the areas analyzed. For this assignment, locations of the mining centers and size of the surrounding population were obtained. Coordinates for existing and future mine projects, rivers, and infrastructure were extracted and acquired from SRK's public data library, the Internet (particularly the African Development Bank Group website), and the Africa Power–Mining Database 2014. Most of the population data, as well as other infrastructure data, were based on World Bank information, including the Africa Infrastructure Country Diagnostic.

World datasets of the Environmental Systems Research Institute (ESRI) were used as the base-data layers within the maps (country boundaries and towns). Such layers as mine project sites, electricity grids, other infrastructure, and rivers were then overlaid.

Generation cost. Power generation costs were estimated using the Model for Electricity Technology Assessment (META model) that ECA developed for the Energy Sector Management Assistant Program (ESMAP). The unit generation costs are defined as the sum of operation and maintenance cost ($ per year) and levelized capital cost ($ per year) divided by net electricity generated per year (kilowatt-hour [kWh] per year):

$$C_{UG} = \frac{(C_{O\&M} + A)}{(GE_G - Au_C)}$$

Where
C_{UG} = unit generation cost ($ per kWh),
$C_{O\&M}$ = operating and maintenance costs ($ per kWh),
A = levelized generation capital costs ($ per kWh),
GE_G = gross electricity generated per year (kWh per year) (capacity [kW] * load factor (%) * 8,760), and
Au_C = auxiliary power consumed per year (kWh per year) (GE_G [kWh per year] * auxiliary power ratio [%]).

The levelized generation capital cost represents an annualized capital cost of power station. Levelized capital cost was calculated using the following equation:

$$A = B * \frac{r * (1+r)^n}{(1+r)^n - 1}$$

where
A = levelized capital cost for each year,
B = capital investment,
n = plant life, and
r = discount rate.

The operating and maintenance (O&M) costs per year ($C_{O\&M}$) were calculated using the following equation:

$$C_{O\&M} = C_{F-O\&M} + C_{n-fV} + C_F$$

where
$C_{O\&M}$ = O&M costs ($ per year),
$C_{F-O\&M}$ = fixed O&M costs ($ per year),
C_{n-fV} = non-fuel variable O&M costs ($ per year), and
C_F = fuel costs ($ per year).

The Power of the Mine • http://dx.doi.org/10.1596/978-1-4648-0292-8

The main parameters that affect the above calculations are as follows:

- Power-plant capacity, which will affect capital and O&M costs; power plants with higher capacities usually have lower levelized capital costs and lower O&M costs
- Fuel price and fuel heat content
- Carbon price
- Load factor
- Power plant efficiency
- Discount factor

Transmission cost. Transmission constraints are a major reason for lack of integration between power and mining in various countries. A transmission costing tool was developed to estimate transmission line tariffs that would fully recover whole lifetime costs, with the following features:

- Different parameters were applied in each of the African power pools (West Africa, East Africa, and southern Africa) to reflect regional costs and the technical characteristics of each.
- Capital and operating spending was estimated from recent analogous projects.
- Correction factors for the terrain were applied to calculate specific costs for each project individually.
- Possible connections with existing transmission lines were identified to calculate the costs for convertor stations.
- Losses on the transmissions lines were allowed for.
- Financing terms were included (debt financing costs and return-on-investor equity).
- The result was tariffs that would fully recover whole lifetime costs, calculated on a net-present-value basis.

This transmission costing tool only provides an initial, high-level estimate of the costs of a new transmission line. Before a transmission project can be properly formulated, load flow and network stability studies would need to be conducted; these could reveal the need for different design options or additional investments in control and protection equipment.

Capital costs cover all elements of the construction of the power transmission line. They include the costs for materials, components, installation, reactive power compensators, project management, and contingency. To identify the capital costs that apply for each power transmission technology, the transmission lines are categorized by voltage level (I = 132 kilovolt (kV), 220 kV, 400 kV and so on), current flow (j = AC or DC), and type of circuit (k = single, double, or bipolar). It should be noted that capital costs vary by time and place due to changes in material prices and labor costs; however, a standard set of cost assumptions are used for all case study countries. The construction period was assumed to be 2 years with an equal spread of capital costs between years 1 and 2.

O&M cost is assumed to be 3 percent of capital cost.

Economic benefits for mines, population, and environment. The economic benefit for the mines was calculated as follows:

$$EB_m = (CPP - TC - SSC) * C_m$$

where
EB_m = economic benefit of the mines ($),
CPP = centralized power plant costs ($ per megawatt-hour [MWh]),
TC = transmission costs ($ per MWh),
SSC = self-supply costs ($ per MWh), and
C_m = present value of total mining consumption (MWh).

The economic benefit for the population compared the sum of generation and transmission unit costs with the population's willingness to pay (measured as cost of dry-cell battery expenses), calculated as follows:

$$EB_p = (WTP - CPP - TC) * C_p$$

where
EB_p = economic benefit for the population ($),
WTP = willingness to pay for the population ($),
CPP = centralized power plant costs ($/MWh),
TC = transmission costs ($/MWh), and
C_p = present value of total residential and nonresidential consumption (MWh).

All values were calculated in present value terms for the 20-year period of the study.

The environmental benefit was estimated using the fuels emissions factor and heat rates of the power plants and estimated consumption volumes; it was calculated as follows:

$$em = ef_i * HR_j * C_m$$

where
em = kilograms of carbon dioxide ($kgCO_2$) of emission,
ef_i = emissions factor of fuel i ($kgCO_2$ per GJ),
HR_j = heat rate of power plant j (gigajoules [GJ] per MWh), and
C_m = present value of total mining consumption (MWh).

Reference

Santley, David, Robert Schlotterer, and Anton Eberhard. 2013. *Harnessing African Gas for African Power.* Washington, DC: World Bank.

Environmental Benefits Statement

The World Bank Group is committed to reducing its environmental footprint. In support of this commitment, the Publishing and Knowledge Division leverages electronic publishing options and print-on-demand technology, which is located in regional hubs worldwide. Together, these initiatives enable print runs to be lowered and shipping distances decreased, resulting in reduced paper consumption, chemical use, greenhouse gas emissions, and waste.

The Publishing and Knowledge Division follows the recommended standards for paper use set by the Green Press Initiative. Whenever possible, books are printed on 50 percent to 100 percent postconsumer recycled paper, and at least 50 percent of the fiber in our book paper is either unbleached or bleached using Totally Chlorine Free (TCF), Processed Chlorine Free (PCF), or Enhanced Elemental Chlorine Free (EECF) processes.

More information about the Bank's environmental philosophy can be found at http://crinfo.worldbank.org/wbcrinfo/node/4.

www.ingramcontent.com/pod-product-compliance
Lightning Source LLC
Chambersburg PA
CBHW082356270326
41935CB00013B/1644